Forced Movements, Tropisms, and Animal Conduct

BY

JACQUES LOEB

With a New Introduction by
JERRY HIRSCH
PROFESSOR OF PSYCHOLOGY AND ZOOLOGY
UNIVERSITY OF ILLINOIS

DOVER PUBLICATIONS, INC.
NEW YORK

Copyright © 1973 by Dover Publications, Inc.
All rights reserved under Pan American and
International Copyright Conventions.

This Dover edition, first published in 1973, is an unabridged and unaltered republication of the work originally published by J. B. Lippincott Company, Philadelphia, in 1918. A new Introduction has been written by Jerry Hirsch especially for the Dover edition.

International Standard Book Number: 0-486-60989-8
Library of Congress Catalog Card Number: 72-92761

Manufactured in the United States of America
Dover Publications, Inc.
180 Varick Street
New York, N. Y. 10014

"It would . . . be a misconception to speak of tropisms as of reflexes, since tropisms are reactions of the organism as a whole, while reflexes are reactions of isolated segments" (p. 23).

"Tropisms, of course, are now a well-accepted part of experimental biology" (Northrop 1961, p. 3).

INTRODUCTION TO THE DOVER EDITION

THIS book appeared late in October 1918, shortly before the armistice to World War I *(Publisher's Weekly,* 1918, *94,* pp. 1318, 1342). It represents the last and most complete treatment by Jacques Loeb (1859–1924) of the problems of behavior. By then, Loeb had become deeply committed to research in colloid chemistry, the forerunner of today's protein chemistry and molecular biology. Ironically its publication coincided with the beginning of an undeserved eclipse in America of Loeb's *acknowledged* influence on behavior study, and the consequent failure to appreciate his contribution almost deprived behavioral biology of one of its fundamental categories.

OBJECTIVISM

While, in fact, he was anticipated by T. H. Huxley's cogent analysis of the irrelevance of imputing consciousness to the crayfish (Huxley 1874, p. 219; Piéron 1941, p. 4),[1] it

---

[1] In this Introduction references appear in several forms: in parentheses a number either by itself or followed by a page number refers to a numbered reference in Loeb's Literature list at the end of his text. A page or chapter number in parentheses refers to that part of Loeb's text thus indicated. Other references are to the Literature list following this Introduction.

was Loeb in 1888 (285) and 1889 (287) who established and proclaimed the importance of the objective study of behavior and who decried the futility of subjectivism and anthropomorphism, not Watson who in 1913 merely attached the label behaviorism to that point of view but now is accorded the credit for what it represented—a priority fully appreciated by Blum (1954, p. 320). As I have previously pointed out (Hirsch 1967a, p. 123), Watson described having been a student under Loeb at the University of Chicago (hence the influence) :

> ... I took biology, and then physiology under Loeb. Loeb wanted me to do my research under him. ... Neither Angell nor Donaldson ... felt that Loeb was ... "safe" ... for a green Ph.D. candidate, so I took my research jointly under Donaldson and Angell. ... My debt to ... Loeb can never be repaid. ... I learned ... from Loeb the fact that all research need not be uninteresting (Watson 1936, pp. 273, 275).

Since we have so often ridiculed the propensity of our Russian contemporaries for claiming undeserved historical priority for too many developments, I include the apt account by perhaps the best-informed psychologist of this century, Henri Piéron, who edited *l'Année Psychologique* for over 50 years beginning about 1912:

> ... I had renounced the subjective study of the phenomena of consciousness and ... affirmed the validity of the biological science of behavior for man and animals well before the spread of that psychological "Behaviorism" which Watson claimed as uniquely American and about which the only thing unique was his often puerile exaggerations (Piéron 1958, p. viii, my translation).

Even the arch-opponent of the tropism theory, H. S. Jennings, had appreciated and acknowledged as early as 1908:

... the tremendous service done by Loeb in championing through thick and thin the necessity for the use of objective, experimental factors in the analysis of behavior. ... the clearcut enunciation, defense and application of the principles on which valuable experimental work has rested since that time, and on which it must continue to rest (Jennings 1908, p. 708).

Furthermore, Watson also reports:

In the Fall of 1908 I went to Hopkins. ... As soon as possible, I began work with Jennings, taking his courses in evolution and his lab on the behavior of lower organisms (Watson 1936, p. 276).

Thus Watson was in continuous contact with Jennings at the very time Loeb's incontrovertible priority was being established in America. Several of the more recent historical discussions (Burnham 1968; Littman 1971; MacKenzie 1972) have omitted this information.

Two developments seem to have contributed to Loeb's eclipse. First, with the advent of American behaviorism there came a disenchantment with the concept of instinct and a shift of emphasis to that of learning. Of course, Loeb had voiced his own criticisms of the unjustified teleology he found in contemporary usage of instinct (see Chap. XVIII). Second, there was the appearance in 1919 of Alfred Kühn's *Die Orientierung der Tiere im Raum* (The orientation of animals in space). It provided quite a different and completely *descriptive* (theoretically neutral about possible mechanisms in behavior) framework, which was to become the basis for Fraenkel's and Gunn's (1940) *The orientation of animals: kineses, taxes and compass reactions*. The latter is now described as "a summary and interpretation of all previous important experimental work, from the day of Loeb to the present. It is the only comprehensive treatment of the subject in existence, and has become the universally accepted guide to all who deal with the subject—whether biologists or psychologists." Its account of orientation behavior (as advertised by the previous quotation from its cover) has been almost universally (and uncritically, see Hirsch 1967b, p. 395) accepted, quoted and paraphrased by biology and psychology texts. As the authors state in the preface to its 1961 Dover reprinting:

> today hardly a single zoologist will use the word *tropism* for behaviour of motile animals; indeed the tendency in the last two decades has been to go even farther than we have done away from Loeb's ideas (Fraenkel and Gunn 1961, pp. v-vi).

—a statement to be contrasted with Northrop's in the epigraph to this Introduction.

For Loeb the tropism concept provided an antidote to two tendencies impeding the establishment of a scientifically valid behavioral biology as well as describing a fundamental category of behavior: reactions of the organism as a whole (p. 23). One was the anthropomorphic attribution of free will to even the most primitive organisms. As Brett's *History* (1921, p. 303) records:

> The kindly observers who from 1860 to 1890 entertained a large public with curious narratives were rudely silenced by the reports which Jacques Loeb published in the last year of that epoch. From this work arose a new type of comparative psychology, the mechanistic school.

Loeb left no doubt about his intentions:

> . . . the history of science has taught us that confusion always reigns when anthropomorphic motives are brought into scientific research. Before the time of Galileo a body sinking in a fluid "sought its place" (see Mach, *Geschichte der Mechanik*, 1st ed., p. 117). Galileo and his followers put an end to the sovereignty of this psychology, at least in inanimate nature. Mankind has had no reason to regret

this revolution. In biology, however, even at this date, protoplasmic substances still move toward the source of light "because of curiosity" (294, p. 81).

The other tendency was the glib use of natural selection to explain too much in biology, a situation analogous to the labeling of instincts in psychology against which, following Loeb, the behaviorists were to rebel. Both tendencies discouraged mechanistic analysis—the only kind from which scientific knowledge can grow.

TROPISMS

Two tropism studies published before Loeb emigrated in 1891 from Europe to America have proved to be of fundamental importance. Using the crab, *Cuma Rathkii*, which lives in mud beneath the harbor at Kiel, Germany, Loeb described a response not readily attributable to the action of natural selection—a non-adaptive tropism. When placed in an aquarium and stimulated by directional light, the crabs moved toward the light source even though their natural habitat does not offer them the opportunity to behave in this way. Loeb also showed in the same experiment that, when provided with mud in the aquarium and given an opportunity, this crab ignored the light and buried itself in the mud. Loeb never denied evolution. He objected to its misuse. In his address "The significance of tropisms for psychology" presented to the VIth International Congress of Psychology, Geneva, 1909, he explained:

> We must, therefore, free ourselves at once from the overvaluation of natural selection and accept the consequences of Mendel's theory of heredity, according to which the animal is to be looked upon as an aggregate of independent hereditary qualities (Loeb 1910, p. 53).

Apparently the crab's mud-burrowing response is an adaptation overriding its powerful and non-adaptive phototropism. The other important study (p. 63) showed how the sessile marine annelid *Spirographis spallanzani* reacted to light in the same way as did the plants studied by Loeb's friend and unofficial mentor, the renowned German botanist, Julius von Sachs. These then were the "proofs" of "The heliotropism of animals and its identity with the heliotropism of plants" (287, 288, translated and republished in 294). His point being: since we do not attribute will, desire and mind to plants, there is no reason to attribute such properties to animals, which perform in the same way under the same conditions.

Loeb envisioned a biology that would become as completely deterministic as he believed physical science to be. Under the right conditions, the behavior of an organism—plant, animal or human—is completely determined and therefore accurately predictable. It will only be by studying organisms under well-controlled conditions that we shall come to understand the nature of their organization.

In 1936 Ullyott reported that planaria, when placed in a horizontally nondirectional gradient of vertical overhead light, aggregate (literally are trapped) at the dark end of the gradient, *because* their random turning movements occur more rapidly in brighter light and more slowly in dimmer light. It was claimed to be the increased activity at the bright end of the gradient and the decreased activity at the dim end which carried them away from the bright region and trapped them into aggregating in the dim region, *not* the directionality of the stimulating light. Ullyott believed he had eliminated directional horizontal light by masking all reflecting surfaces with dead black paint. This "phenomenon" reported by Ullyott (1936*a*, *b*) became the prototype of klino-kinesis, which Fraenkel and Gunn then combined with Kühn's system into the descriptive framework that has been so widely accepted since it appeared in 1940.

INTRODUCTION TO THE DOVER EDITION   xi

When Koehler criticized and called attention to the limitations to their system—a task previously undertaken by Blum (1935; Blum, Hyman and Burdon 1936)—Gunn replied with the following plea:

> Finally may I appeal once more to workers in behaviour not to take part in the pointless destruction of a limited but useful system in this merely incidental way, not to begin again fruitless discussion of whether any behaviour is forced or free, but to study behaviour as a subject in its own right, without depending upon doubtful generalizations from more elemental subjects (Gunn 1950, p. 303).

It was to require another sixteen years for Gunn (who wrote the klino-kinesis discussions, see Preface to First Edition, Fraenkel and Gunn 1961, p. ix) to voice his own doubts:

> ... the scheme (klino-kinesis) ... depended upon only two papers by Ullyott on *Dendrocoelum lacteum* and several attempts to repeat this work have failed. ... so this class of reaction is most insecurely based. The word "klino-kinesis" is freely used but nobody really knows what it means; let us hope that someone will tackle the matter thoroughly (Gunn 1966, p. 110).

In 1968 at the University of Toronto, Aivars Stasko completed as his doctoral dissertation an extensive study attempting to replicate the evidence for klino-kinesis. When scattered lateral (horizontal) light was truly eliminated, no orientation or aggregation occurred. Whenever orientation did occur, it was found to be correlated with the presence of directional light. This study has now been published as a monograph co-authored by Charlotte M. Sullivan, Stasko's sponsor and collaborator. They report: "In conclusion, the results ... leave Ullyott's hypothesis without a vestige of support" (Stasko and Sullivan 1971, p. 110).

In our laboratory over the past five years, first Gerald

Pond, then Steve A. Platt, and now Daniel Wolf and I have replicated and extended certain very important observations reported by Gaston Viaud, who began working at the University of Strasbourg about 1930 (Hirsch 1967*b*, *c*; Platt 1972, Platt and Hirsch 1968, 1970, and in preparation). There now exists substantial quantitative evidence for the forced-movement nature of galvanotropism in the goldfish *Carassius auratus*. In low-density direct electric current the fish are literally drawn (forced!) to the cathode, or negative electrode. Furthermore, at current offset they neither avoid the region of the cathode (it's not aversive) nor do they seek it (their approach has not been conditioned). Our picture of galvanotropism in the intact organism is consistent with that for the intense galvanotropism shown by parts of organisms, which made Marsh's planarian regeneration studies so difficult (Marsh and Beams 1952, p. 192).

Gaston Viaud proposed a resolution for the so long unresolved and bitter controversy between Jacques Loeb and H. S. Jennings—between the interpretation of certain animal behaviors as being either forced, often non-adaptive, movements controlled and directed by outside sources of physical energy, or the highly adaptive result of trial and error, a selection of successful responses by the individual and a selection of successfully responding individuals (to pass on genes) by the species. Unfortunately Viaud's important work is not well enough known to those not intimate with the French literature. The bibliography of his contributions from 1930 until his death on 29 December 1961, conveniently assembled by his student Jean Médioni (1962), lists over 100 publications, largely devoted to the detailed (and *extremely valuable*) quantitative experimental analysis of tropisms—first, phototropism in daphnia, rotifers, copepods, planaria and pigeons; later, galvanotropism in planaria, paramecia, frog tadpoles, and goldfish.

The solution that emerges from Viaud's painstaking work is of fundamental importance for *any* theory of behavior. Briefly, what Viaud showed is that both Loeb and Jennings were wrong in part of what they each asserted, and that both of them were correct in another part. They both erred in failing to distinguish between adience and abience, between going toward a source of (electromagnetic energy) stimulation and going away. For Loeb both adience and abience were non-adaptive forced movements—his tropisms. He made no essential distinction between approach and avoidance. Ironically, Jennings made a comparable mistake. For him both adience and abience were adaptive behavior—his trial and error.

Conceptually, Viaud's method and solution were both extremely simple, just as was Mendel's only somewhat more solitary treatment of another important problem. Like Mendel's solution to the enigma of heredity, Viaud's solution required a large number of observations entailing much hard work. Instead of merely presenting a stimulus, such as a directional source of light, to an animal and then recording whether it responds photopositively (like the proverbial moth at a candle) or photonegatively (like the cockroach fleeing light)—experimental observations requiring only a brief moment to record (analogous to a snapshot)—Viaud observed the reactions of his animals to a single stimulus intensity over extended periods of time—anywhere from one-half hour to two hours under controlled laboratory conditions. Then he repeated similar observations at each of several stimulus intensities, i.e., he performed an extended psychophysical analysis. From the perspective of his more thorough analysis Viaud found that his several species were *both* photopositive and photonegative, then he found the species studied later to be both galvanopositive (drawn to the cathode) and galvanonegative (driven away from it)—behavior is polyphasic. Furthermore, his analysis showed the characteristics of the movements toward (adience) or away from (abience) the source of stimulation to be quite different. The curve for the speed of approach to light as a function of intensity approximates the Weber-Fechner ogive and is consistent with what Viaud was to call, in agreement with Georges Bohn (1933, 1940), a tropism or polarized kinesis—a forced non-adaptive movement, oriented and driven by the stimulus as Loeb had claimed. The analogous curve for the speed of movement away from light as a function of intensity (and time) is quite different. Over a considerable range of intensity (for a given time interval) speed of movement changes little until a threshold is passed, intensity becomes too great, and a sudden avoiding or flight reaction is "released" (triggered)—clearly an adaptive response as Jennings had claimed.

For quite some time it has been incorrectly assumed that Jennings succeeded in using "the concept of trial and error . . . . to demolish Loeb's doctrine of animal tropisms" (Fleming 1964, p. xxxvi). While that belief did attain wide currency, it has now become clear that Loeb's work has far more to say to us today than has been appreciated by all but a few. Grassé (1961, p. 2814) reports that l'Académie des Sciences de France awarded Viaud the Montyon Prize in recognition of his unique contribution. It was Grassé, with the obvious concurrence of Piéron, who chose Viaud to make the lead presentation (on tropisms!) at the important Instinct Colloquium in Paris (Grassé 1956, p. 789). There is moreover an enormous amount of well-substantiated evidence for "the trapping effect" of light (Verheijen 1958).

Thus the concept of a tropism as a polarized kinesis (the interaction of stimulus polarity and organismic structure) is seen to be of fundamental importance for behavior theory. Without it there is no way to explain the most primitive

INTRODUCTION TO THE DOVER EDITION xv

feature of attention: why any organism (animal or plant!) ever focuses on (orients toward) any stimulus in the first instance. *Given* some initial stimulus orientation, conditioning or learning theory can be invoked to explain how the same individual might come to respond (orient) to subsequently appearing stimuli, because of their reinforcement by or association with the initial stimulus, etc. Again, *given* some initial stimulus orientation, evolutionary theory (natural selection) can be invoked to explain how a species might come to respond (orient) to an entire series of stimuli associated with the initial one, because of the adaptive value in the organized behavior patterns that evolve (ritualization) as stimulus-response chains, e.g., courtship, parental behavior, and so forth (Hirsch 1971).

An objection has also been raised that

> Loeb [312], having an uncanny knack for choosing organisms that "fit" his hypotheses, tested the validity of the Bunsen-Roscoe Law for the heliotropic movement of the hydroid *Eudendrium*. . . . [because his results fit theory admirably, the critic continues]. Loeb concluded that the Bunsen-Roscoe Law was valid for the heliotropic response of *Eudendrium*. Unfortunately, however, he overreached himself in interpreting his experiment; he speculated that the experiment . . . and others like it, showed "that the tropism theory not only allows us to predict the nature of animal reactions but also allows us to predict them quantitatively. Thus far the tropism theory is the only one which satisfies this demand of exact science (p. 93)."
>
> It is not difficult, after reading the passages quoted above, to understand why Jacques Loeb did not want for critics (Gussin 1963, pp. 325–26).

Finding the right organism and the appropriate conditions for analyzing fundamental relations is a virtue, not a sin; its importance can hardly be overemphasized. After working out the fundamental relations in heredity, Mendel—that amateur who had the good sense to experiment on the ordinary garden pea plant *Pisum sativum*—went on, with the encouragement of Naegeli the professional, to work with hawkweeds of the genus *Hieracium*. The subsequent experiments "failed," because they did not replicate the results obtained with *Pisum*. What Mendel could not know and what did not become clear until later is that *Hieracium* have a strong tendency to apomixis (a plant analogue of parthenogenesis or development of unfertilized ovules, thus genetic segregation is absent, Iltis 1932, Chap. 12; Olby 1966, p. 89) and some species might be polyploid (Darlington 1963, p. 59). In contrast to Gussin's cavil, Osterhout has observed that the questions Loeb

> ... put to nature were never dull and in consequence the answers he received were always interesting, sometimes startlingly so. He did not begin to work until he felt that he had framed the question properly. ... [and] he recognized that the chance of a successful answer lay largely in the choice of material. In this respect he displayed great sagacity. It is said that when he began his work on tropisms he was found among the cases of the museum looking for animals that most resembled plants. True or not, the story illustrates his habit of mind (Osterhout 1928, p. 362).

Impact and Personality

Loeb's lifetime reputation and posthumous eclipse now pose a recognized historical puzzle:

> Given Loeb's contemporary eminence and the many evidences of the esteem in which he was held by his fellow scientists, the historian is faced with the delicate problems of explaining his later neglect by scholars and of determining whether it was justified. Whatever the final verdict may be, there is much evidence in the papers [at the Library of Congress] that Loeb merits careful study by both historians of science and by students of American intellectual history (Reingold 1962, p. 119).

Certainly both professionally and personally perhaps no other scientist of the period attained greater eminence if we take for the measure of eminence the recognition, respect and affection expressed by one's colleagues the world over. Between 1901 and his death on 11 February 1924 about one hundred recommendations from ten countries nominated him for the Nobel prize (Nobel Foundation 1962, p. 256). Between 1901 and 1925, however, ten of the fifty prizes scheduled for Chemistry and Physiology or Medicine, the two fields in which Loeb obviously qualified, i.e., 20 per cent, were awarded to no one! Though denied to Loeb, the prize was awarded in 1946 to John H. Northrop, his student and collaborator (Loeb and Northrop 1916*a*, *b*; 1917*a*, *b*, *c*; Northrop and Loeb 1922–1923).

Northrop had a deep appreciation of Loeb's merit and this was expressed in his obituary:

> It was in devising and carrying out experimental tests of his ideas that his genius reached its greatest heights. He did not believe in conquering an obstacle by laying siege with laborious and painstaking experiments, but preferred to spring upon it from some totally unexpected angle and decide the issue with a simple but marvelously ingenious experiment. To the chemist this method was at times disconcerting. A reviewer of one of his books remarked, "Some of the methods would curdle the blood of an analytical chemist." They furnished results which answered the question, however, and that was to him the essential point. Like Faraday, he had at times the uncanny gift of knowing the truth before the work was done. He would sometimes say "I know what it is. The question is, how to prove it." And yet a single trustworthy experiment would cause him to give up at once a theory he had upheld with the greatest enthusiasm (Northrop 1924, p. 318).

Loeb's attitude and approach to science were perhaps

best articulated by his own words in his last publication:

> We are already in possession of an enormous number of enigmatic though often interesting observations . . . in different animals and plants, and it seemed of little value to add to this store of riddles. It is primarily not more facts which are needed . . . but a method and a principle which allow us to pass from the stage of blind empiricism to the stage of oriented search. As long as the investigation of a natural phenomenon is in the stage of blind empiricism we never know what to look for in our experiments nor what to measure, and we are not able to judge whether we are on the road to progress or whether we are losing ourselves in a jungle of futile experiments. With a rationalistic law and a rationalistic method as guides, this danger is avoided (1924a, p. vii).

With respect to Loeb's place in the history of science, it is most significant to find that neither the perspective of 37 years hindsight nor his own ostensibly greater success was to erode Northrop's respect and evaluation. He devoted about a tenth of his autobiographical essay to a further appreciation:

> Loeb. . . . proved that the behavior of many plants and animals could be accurately predicted. . . . This theory of tropisms was attacked largely by anthropomorphic criticisms. . . . Tropisms, of course, are now a well-accepted part of experimental biology.
> Loeb . . . planned and carried out his extraordinary experiments on artificial parthenogenesis, experiments that for originality of conception and brilliancy of performance have rarely been surpassed. They are still, I believe, the nearest approach to the creation of life.
> At the time I entered the laboratory, he was about to attack the problem of the peculiar properties of proteins—in particular their osmotic pressure, viscosity, and swelling. Proteins at that time were considered to be "colloidal complexes" of no definite molecular weight, which combined with electrolytes by a process of adsorption. This explana-

tion . . . contained unnecessary assumptions. . . . Loeb was able, with the help of Donnan's theory of membrane equilibrium, to explain all the complicated phenomena of osmotic pressure, viscosity, and swelling [Loeb 1924b]. This explanation rests on firmly established chemical and physical theory. . . . These results founded the modern theory of protein chemistry (Northrop 1961, pp. 3-4).

One of psychology's most competent scientists also appreciated Loeb and saw how fundamental to the understanding of behavior is the approach articulated in his tropism theory. Shortly before his death, Clarence H. Graham informed me (private communication, 14 April 1971) that he had just written in his autobiography:

In 1927 . . . at Clark . . . . I soon came under Hunter's influence, especially through his seminars Animal Behavior and Principles of Psychology. . . . Although I responded positively to Hunter's statements, I remember that on one matter I made a choice of systematic approach that neither he nor the other students . . . supported. In our study of the points of view represented by Jennings and Loeb, I favored Loeb. I have since recognized . . . that in fact Jennings's approach would probably be favored by many or most students of behavior, but even then I had the feeling that for me an account of behavior would require some more explicitly analytic and testable variables as objectives of study. And so, it is probably not surprising that my interests in sensation and perception have been along the paths laid out by such workers as Helmholtz, Maxwell, Hering, Mach and, to repeat, Loeb (Graham, in press).

Loeb was both an idealist and "mildly sympathetic toward the Socialists" (Reingold 1962, p. 127). He was born in the Rhine Province of Prussia on 7 April 1859 at Mayen near Coblenz, in a region in which French and German culture had long been mixed. While his schooling was in German, it has been said (Robertson 1926–1927) that French was the language of his father's house where he was brought up on the literature of the French Revolution. The influence of that literature upon his outlook remained unabated throughout his life and was acknowledged when he dedicated his book, *The organism as a whole*:

> ... to that group of freethinkers, including D'Alembert, Diderot, Holbach, and Voltaire, who first dared to follow the consequences of a mechanistic science—incomplete as it then was—to the rules of human conduct and who thereby laid the foundation of that spirit of tolerance, justice, and gentleness which was the hope of our civilization until it was buried under the wave of homicidal emotion which has swept through the world (Loeb 1916, p. viii).

The outbreak of World War I had shocked him deeply.

In 1921 Loeb sent a copy of the present book to United States Supreme Court Justice Louis D. Brandeis, and wrote him in a letter:

> I think one day—by some future generation [the ideas put forth in my book]—may be elaborated into a mathematical theory of human conduct (Gussin 1963, p. 321).

He firmly believed that *ultimately* mechanistic biology would explain human behavior, but, as indicated in his reply to the novelist Theodore Dreiser's inquiry about such a possibility, it was not to be expected in the foreseeable future (Reingold 1962, p. 129). Still, in the spirit of the Enlightenment, he remained optimistic that when it does happen it will be for human good.

As a scientist he has often been described as an indefatigable worker

> ... unresting, tireless.... The secret of his productivity as an investigator lay in his immense industry, and in the extraordinary sagacity with which he designed his experiments to elicit an unequivocal answer to the questions he had in mind (Robertson 1926-1927, p. 127).

Remarkable also was

INTRODUCTION TO THE DOVER EDITION xxi

> ... the exacting and intelligent but meticulous care he used in repeating and checking all his observations before publication. ... few physicists with the possible exception of Rutherford ... made as sure of the accuracy of experimental data published as did Jacques Loeb (Loeb 1959, p. 4).

As a teacher he was found to be "inspiring, as a lecturer at once brilliant and easy to understand" (Robertson 1926–1927, p. 126).

Recently on our campus the distinguished chemist (two-term member [1964–1976] and chairman since 1970 of the National Science Board) Herbert E. Carter stated on retiring early (age 60) after 40 years at the University of Illinois (his last four as Vice-Chancellor): "Any professor in a good department who gives up his job to go into administration is a damn fool" (Peters 1971). Therefore, to close this glimpse of Loeb the man, I quote again from the sensitive appreciation of his life and work written by an associate for five years at the University of California, Berkeley:

> He did not share that love for administrative work for its own sake which the majority of Americans possess in such a high degree, nor did he relish attendance upon the perennial committee-meetings in which the academic soul delights. He was an uncomfortable member of a committee, because he always insisted upon dragging fundamental principles into the discussion. There is nothing which so shocks your good committee-man as a naked first principle. He hastens to clothe it in a garment of compromise, no matter how patched and threadbare the improvised covering may be. Promptly Loeb tore it off again, once more revealing the principle in all its utter truth and simplicity, a proceeding which did not invariably tend towards the production of harmony (Robertson 1926-1927, p. 126).

So today, half a century after Loeb, we see that *plus ça change plus c'est la même chose.*

We are indebted to Professor Gottfried S. Fraenkel for recommending reprinting this volume and that I write this discussion.

The research was supported by National Institute of Mental Health Training Grants 5 TO1 MH 10715 BLS (06 and 07) for Research Training in the Biological Sciences.

LITERATURE

Three types of references appear in this list of Literature: (1) publications referred to in the Introduction, (2) reviews of this book (herein called Review), and (3) obituaries and some other useful discussions.

ANONYMOUS: (Review). *The Dial*, 1919, *66*, 428, 430.

ARMSTRONG, H. E.: A chemist's homage to the work of a biologist. *Journal of General Physiology*, 1928, *8*, 653–670.

BERNARD, L. L.: (Review). *American Journal of Sociology*, 1919, *25*, 240–241.

BLUM, H. F.: An analysis of oriented movements of animals in light fields. *Photochemical reactions,* Cold Spring Harbor Symposia on Quantitative Biology. The Biological Laboratory, Cold Spring Harbor, L. I., N. Y., 1935, *III*, 210–223.

BLUM, H. F.: Photoorientation and the "tropism theory." *Quarterly Review of Biology*, 1954, *29*, 307–321.

BLUM, H. F., HYMAN, E. J., and BURDON, P.: Studies of oriented movements of animals in light fields. *University of California Publications in Physiology*, 1936, *8*, 107–118.

BOHN, G.: Jacques Loeb (1859–1924). *Comptes Rendus de la Société de Biologie*, 1924, *90*, 728–730.

BOHN, G.: Tropismes. *Traité de Physiologie Normale et Pathologique*, 1933, *9*, 213–234.

BOHN, G.: *Actions directrices de la lumière*. Paris: Gauthier-Villars, 1940.

BRETT, G. S.: *A history of psychology*. New York: The Macmillan Company, 1921.

BUDDENBROCK, W. VON: A criticism of the tropism theory of Jacques Loeb. *Journal of Animal Behavior*, 1916, *6*, 341–366.

INTRODUCTION TO THE DOVER EDITION xxiii

BURNHAM, J. C.: On the origins of behaviorism. *Journal of the History of the Behavioral Sciences*, 1968, *4*, 143–151.

CHILD, C. M.: (Review). The organism as a whole by Jacques Loeb. *Botanical Gazette*, 1918, *65*, 274–280.

CORNER, G. W.: *A history of the Rockefeller Institute, 1901–1953: origins and growth.* New York: The Rockefeller Institute Press, 1964.

CRAIG, W.: (Review). *The Psychological Bulletin*, 1919, *16*, 151-159.

CROZIER, W. J.: (Review). *Science*, 1919, *49*, 171–172.

DARLINGTON, C. D.: *Chromosome botany and the origins of cultivated plants.* London: Allen & Unwin, Ltd., 1963.

DEKRUIF, P. H.: Jacques Loeb, the mechanist. *Harper's Monthly Magazine*, 1923, *146*, 182–190.

DUFFUS, R. L.: Jacques Loeb: mechanist. *The Century*, 1924, *86*, 374–382.

FLEMING, D.: See J. Loeb 1964.

FLEXNER, S.: Jacques Loeb and his period. *Science*, 1927, *66*, 333–337.

FRAENKEL, G. S., and GUNN, D. L.: *The orientation of animals: kineses, taxes and compass reactions.* New York: Dover Publications, Inc. 1961. Originally published in 1940.

GRAHAM, C. H.: In G. Lindzey (ed.): *A history of psychology in autobiography vol. VI.* New York: Appleton-Century-Crofts, in press.

GRASSÉ, P.-P. (organisateur): *L'Instinct dans le comportement des animaux et de l'homme.* Paris: Masson et cie Éditeurs, 1956.

GRASSÉ, P.-P.: Séance du 9 Décembre 1961, physiologie. *Comptes Rendus Académie des Sciences*, 1961, *253*, 2814.

GUNN, D. L.: Discussion in J. F. Danielli and R. Brown (eds.): Symposia of the Society for Experimental Biology, No. IV: *Physiological mechanisms in animal behaviour.* London: Cambridge University Press, 1950, 303.

GUNN, D. L.: Discussion period in P. T. Haskell (ed.): *Insect behavior, symposium No. 3*, Royal Entomological Society of London, 1966, 109–110.

GUSSIN, A. E. S.: Jacques Loeb: the man and his tropism theory of animal conduct. *Journal of the History of Medicine and Allied Sciences*, 1963, *18*, 321–336.

HIRSCH, J.: Behavior-genetic, or "experimental," analysis: the challenge of science versus the lure of technology. *American Psychologist*, 1967a, *22*, 118–130.
HIRSCH, J.: (Review). Mechanisms of animal behaviour, by P. Marler and W. J. Hamilton, III. *Animal Behaviour*, 1967b, *15*, 394–395.
HIRSCH, J.: Tropisms *are* forced movements. *American Zoologist*, 1967c, *7*, 422.
HIRSCH, J.: Tropisms and attention. *XIIth International Ethological Conference Abstracts*, 1971, 38.
HUNTER, W. S.: (Review). *Psychological Bulletin*, 1919, *16*, 179–180.
HUXLEY, T. H.: *Collected essays, Vol. I: Method and results*. New York: D. Appleton & Co., 1904. Orig. published 1874.
ILTIS, H.: *Gregor Johann Mendel, leben, werk und wirking*. (Originally published 1924.) Translated by E. Paul and C. Paul, *Life of Mendel*, 1932. Reprinted: London: Allen & Unwin, 1966.
JENNINGS, H. S.: The interpretation of the behavior of the lower organisms. *Science*, 1908, *27*, 698-710.
KÜHN, A.: *Die Orientierung der Tiere im Raum*. Jena: Verlag von Gustav Fischer, 1919.
LITTMAN, R. A.: Henri Piéron and French psychology: a comment on Professor Fraisse's note. *Journal of the History of the Behavioral Sciences*, 1971, *7*, 261–268.
LOEB, J.: *The organism as a whole from a physicochemical viewpoint*. New York: G. P. Putnam's Sons, 1916.
LOEB, J.: *Regeneration from a physico-chemical viewpoint*. New York: McGraw-Hill Book Co., 1924a.
LOEB, J.: *Proteins and the theory of colloidal behavior*. New York: McGraw-Hill Book Co., 1924b.
LOEB, J.: The significance of tropisms for psychology (originally in German, 1910, see Osterhout 1928). Reprinted in D. Fleming (ed.): *The mechanistic conception of life*, by Jacques Loeb, Cambridge, Mass.: Harvard University Press, 1964 (see 301).
LOEB, J., and NORTHROP, J. H.: Nutrition and evolution. *Journal of Biological Chemistry*, 1916a, *27*, 309–312.
LOEB, J., and NORTHROP, J. H.: Is there a temperature coefficient for the duration of life? *Proceedings of the National Academy of Sciences*, 1916b, *2*, 456–457.

INTRODUCTION TO THE DOVER EDITION xxv

LOEB, J., and NORTHROP, J. H.: On the influence of food and temperature upon the duration of life. *Journal of Biological Chemistry*, 1917a, *32*, 103–121.

LOEB, J., and NORTHROP, J. H.: What determines the duration of life in metazoa? *Proceedings of the National Academy of Sciences*, 1917b, *3*, 382–386.

LOEB, J., and NORTHROP, J. H.: Heliotropic animals as photometers on the basis of the validity of the Bunsen-Roscoe Law for heliotropic reactions. *Proceedings of the National Academy of Sciences*, 1917c, *3*, 539–544.

LOEB, L.: Autobiographical notes. *Perspectives in Biology and Medicine*, 1958, *2*, 1–23.

LOEB, L. B.: Jacques Loeb: recollections of his career as a scientist. *The Rockefeller Institute Quarterly*, 1959, *3*, 1–4, 6.

MACKENZIE, B. D.: Behaviourism and positivism. *Journal of the History of the Behavioral Sciences*, 1972, *8*, 222–231.

MARSH, G., and BEAMS, H. W.: Electrical control of morphogenesis in regenerating *Dugesia tigrina*. *Journal of Cellular and Comparative Physiology*, 1952, *39*, 191–213.

MÉDIONI, J.: La nouvelle psychologie animale, science biologique. *Cahiers de psychologie*, 1962, *5*, 153–172, 209–216.

NOBEL FOUNDATION (eds.): *Nobel: the man and his prizes.* New York: Elsevier Publishing Company, 1962.

NORTHROP, J. H.: Jacques Loeb (1859-1924). *Industrial and Engineering Chemistry*, 1924, *16*, 318.

NORTHROP, J. H.: Biochemists, biologists, and William of Occam. *Annual Review of Biochemistry*, 1961, *30*, 1–10. Reprinted in *The excitement and fascination of science.* Palo Alto: Annual Reviews, Inc., 1965.

NORTHROP, J. H., and LOEB, J.: The photochemical basis of animal heliotropism. *Journal of General Physiology*, 1922–1923, *5*, 581–595.

OLBY, R. C.: *Origins of Mendelism.* London: Constable & Co., Ltd., 1966.

OSTERHOUT, W. J. V.: Jacques Loeb, the man. *Science*, 1924, *59*, 427–429.

OSTERHOUT, W. J. V.: Biographical memoir of Jacques Loeb, 1859–1924. *Journal of General Physiology*, 1928, Jacques Loeb Memorial Volume, *8*, ix-xcii. And in *National Academy of Sciences Biographical Memoirs*, 1930, *13*, 318–401.

PETERS, P.: After 40 years, a new place in the sun. *Champaign-Urbana Courier*, Sunday, 27 June 1971.

PIÉRON, H.: (Review). *L'Année Psychologique*, 1914–1919, *21*, 299–300.

PIÉRON, H.: *Psychologie Zoologique*. Paris: Presses Universitaires de France, 1941.

PIÉRON, H.: *De l'actinie à l'homme études de psychophysiologie comparée*. Paris: Presses Universitaires de France, 1958.

PLATT, S. A.: *A re-examination of animal tropisms*. Unpublished doctoral dissertation, University of Illinois, 1972.

PLATT, S. A., and HIRSCH, J.: The analysis of (typologically conceived) tropisms. *American Zoologist*, 1968, *8*, 59.

PLATT, S. A., and HIRSCH, J.: Tropisms, electromotive forces, and capacitance. *Animal Behaviour*, 1970, *18*, 399.

REINGOLD, N.: Jacques Loeb, the scientist. *The Library of Congress Quarterly Journal of Current Acquisitions*, 1962, *19*, 119–130.

ROBBINS, W. J.: The influence of Jacques Loeb on the development of plant tissue culture. *Bulletin du Jardin Botanique de l'État*, Brussels, 1957, *27*, 189–197.

ROBERTSON, T. B.: The life and work of a mechanistic philosopher: Jacques Loeb. *Science Progress in the Twentieth Century*, 1926–1927, *21*, 114–129.

STASKO, A. B.: *Klino-kinesis in planarians and their mechanism of selection in illumination gradients of vertical light*. Unpublished doctoral dissertation, University of Toronto, 1968.

STASKO, A. B., and SULLIVAN, C. M.: Responses of planarians to light: an examination of klino-kinesis. *Animal Behaviour Monographs*, 1971, *4*, 45–124.

THOMPSON, D'A. W.: (Review). *Nature*, 1919, *103*, 163–164.

THOMSON, J. A.: (Review). *Scientia*, 1919, *26*, 502–504.

ULLYOTT, P.: The behaviour of *Dendrocoelum lacteum:* I. responses at light and dark boundaries. *Journal of Experimental Biology*, 1936a, *13*, 253–264.

ULLYOTT, P.: The behaviour of *Dendrocoelum lacteum:* II. responses in nondirectional gradients. *Journal of Experimental Biology*, 1936b, *13*, 265–278.

VERHEIJEN, F. J.: The mechanisms of the trapping effect of artificial light sources upon animals. *Archives Néerlandaises de Zoologie*, 1958, *13*, 1–107.

WASHBURN, M. F.: (Review). *The Journal of Philosophy*, 1919, *16*, 554–556.
WATSON, J. B.: In C. Murchison (ed.), *A history of psychology in autobiography vol. III*. Worcester: Clark University Press, 1936, 271-281.
WINDLE, B. C. A.: Loeb and his system. *The Dublin Review*, 1918, *163*, 284–296.

Forced Movements, Tropisms, and Animal Conduct

AUTHOR'S PREFACE

ANIMAL conduct is known to many through the romantic tales of popularizers, through the descriptive work of biological observers, or through the attempts of vitalists to show the inadequacy of physical laws for the explanation of life. Since none of these contributions are based upon quantitative experiments, they have led only to speculations, which are generally of an anthropomorphic or of a purely verbalistic character. It is the aim of this monograph to show that the subject of animal conduct can be treated by the quantitative methods of the physicist, and that these methods lead to the forced movement or tropism theory of animal conduct, which was proposed by the writer thirty years ago, but which has only recently been carried to some degree of completion. Many of the statements, especially those contained in the first four chapters of the book, are familiar to those who have read the writer's former publications, but so much progress has been made in the last few years that a new and full presentation of the subject seemed desirable. Chapters V to XIII and Chapter XVI are partly or entirely based on new experiments.

Only that part of the literature has been considered which contributes to or prepares the way for quantitative experiments.

AUTHOR'S PREFACE

The writer is under obligation for valuable criticism to his wife, to Professor T. H. Morgan, and to Lieutenant Leonard B. Loeb, who were kind enough to read the manuscript.

J. L.

The Rockefeller Institute
for Medical Research,
New York.
March, 1918.

CONTENTS

CHAPTER	PAGE
I. Introduction	13
II. The Symmetry Relations of the Animal Body as the Starting Point for the Theory of Animal Conduct	19
III. Forced Movements	24
IV. Galvanotropism	32
V. Heliotropism. The Influence of One Source of Light	47
1. General Facts	47
2. Direct Proof of the Muscle Tension Theory of Heliotropism in Motile Animals	52
3. Heliotropism of Unicellular Organisms	62
4. Heliotropism of Sessile Animals	63
VI. An Artificial Heliotropic Machine	68
VII. Asymmetrical Animals	70
VIII. Two Sources of Light of Different Intensity	75
IX. The Validity of the Bunsen-Roscoe Law for the Heliotropic Reactions of Animals and Plants	83
X. The Effect of Rapid Changes in Intensity of Light	95
XI. The Relative Heliotropic Efficiency of Light of Different Wave Lengths	100
XII. Change in the Sense of Heliotropism	112
XIII. Geotropism	119
XIV. Forced Movements Caused by Moving Retina Images: Rheotropism: Anemotropism	127
XV. Stereotropism	134
XVI. Chemotropism	139
XVII. Thermotropism	155
XVIII. Instincts	156
XIX. Memory Images and Tropisms	164
Literature	173

ILLUSTRATIONS

FIG.		PAGE
1.	Forced Position of Larva of the Dragon Fly whose Left Cerebral Ganglion is Destroyed....................................	30
2.	Forced Position of Shrimp when Galvanic Current Goes from Head to Tail..	34
3.	Forced Position of Shrimp when Positive Current Goes from Tail to Head...	35
4.	Position of Legs of Shrimp when Current Goes Sideways through the Animal...	37
5–6.	Show Same Effects of Current on the Common Crawfish as Those on Shrimp in Figs. 2 and 3..................................	38
7.	Diagram Indicating the Orientation of the Neurons for Flexor and Extensor Muscles of the Right and Left Legs.................	39
8–9.	Diagram Indicating Orientation of Neurons for Flexor and Extensor Muscles of Third and Fifth Pairs of Legs.....................	40
10.	Forced Position of *Amblystoma* Larva Under Influence of Galvanic Current Going Through Animal from Head to Tail.............	41
11.	Forced Position of *Amblystoma* Larva When Current Goes from Tail to Head..	41
12.	Tentacles and Manubrium of Jellyfish Under Influence of Galvanic Current...	42
13.	Strip of Jellyfish Under Influence of Galvanic Current..........	42
14.	*Paramœcium* Under Normal Conditions........................	43
15.	Current Going Through *Paramœcium*	44
16.	Showing that Positively Heliotropic Animals Will Move from Sunlight into Shade if Illumination of Both Eyes Remains the Same...	50
17.	Position of Water Scorpion When Right Eye is Towards the Light..	53
18.	Positions of *Ranatra* When Light is in Front and Behind Animal....	54
19.	Robber Fly Under Normal Conditions...........................	55
20.	Robber Fly with Right Eye Blackened..........................	56
21.	Position of Robber Fly when Lower Halves of Both Eyes are Blackened	57
22.	Position of Robber Fly when Upper Halves of Both Eyes are Blackened	58

ILLUSTRATIONS

FIG.		PAGE
23.	Diagram Showing Position of the Flagellum as Seen in a Viscid Medium	62
24.	Tube Worms in Aquarium	63
25.	Same Animals After Position of Aquarium was Reversed	64
26.	Polyps of *Eudendrium* all Growing Towards Source of Light	66
27.	Fly with Right Eye Blackened	72
28.	Diagram of Apparatus Used to Produce Differential Bilateral Light Stimulation	76
29.	Diagram to Show Method of Measuring Trails	77
30.	Diagram for Constructing Direction of Motion of Larvæ	80
31.	Method for Proving Validity of Bunson-Roscoe Law	90
32.	A Glass Plate	91
33.	Difference in Gathering Places of Animals	96
34.	Method of Determining the Relative Heliotropic Efficiency of Two Different Parts of the Spectrum	107
35.	Geotropic Curvature of Stems of *Bryophyllum calycinum*	120
36.	All Stems were Originally Straight and Suspended Horizontally	121
37.	When the Size of the Leaf is Reduced by Cutting Out Pieces from the Middle	120
38.	Effect of Cutting off Lateral Parts of the Leaves	121
39.	Influence of Motion of the Hand on a Swarm of Sticklebacks in an Aquarium	132
40.	The Regenerating Polyp of *Tubularia* in Contact with Glass Wall of Aquarium	137
41.	Reactions of *Chilomonas* to a Drop of $\frac{1}{50}$ per cent. HCl	145
42.	Method of Proving the *Paramœcia* are not Positive to Acid of Low Concentration	146

FORCED MOVEMENTS, TROPISMS, AND ANIMAL CONDUCT

CHAPTER I

INTRODUCTION

The analysis of the mechanism of voluntary and instinctive actions of animals which we propose to undertake in this volume is based on the assumption that all these motions are determined by internal or external forces. Our task is facilitated by the fact that the overwhelming majority of organisms have a bilaterally symmetrical structure, *i.e.*, their body is like our own, divided into a right and left half.

The significance of this symmetrical structure lies in the fact that the morphological plane of symmetry of an animal is also its plane of symmetry in physiological or dynamical respect, inasmuch as under normal conditions the tension in symmetrical muscles is the same, and inasmuch as the chemical constitution and the velocity of chemical reactions are the same for symmetrical elements of the surface of the body, *e.g.*, the sense organs.

Normally the processes inducing locomotion are equal in both halves of the central nervous system, and the tension of the symmetrical muscles being equal, the animal moves in as straight a line as the imperfections of its locomotor apparatus permit. If, however, the velocity of chemical reactions in one side of the body, *e.g.*, in one eye of an insect, is increased, the physiological symmetry of both sides of the brain and as a consequence the equality of tension of the symmetrical muscles no longer exist. The muscles connected with the more strongly illuminated eye are thrown into a stronger tension,[a] and if now impulses for locomotion originate in the central nervous system, they will no longer produce an equal response in the symmetrical muscles, but a stronger one in the muscles turning the head and body of the animal to the source of light. The animal will thus be compelled to change the direction of its motion and to turn to the source of light. As soon as the plane of symmetry goes through the source of light, both eyes receive again equal illumination, the tension (or tonus) of symmetrical muscles becomes equal again, and the impulses for locomotion will now produce equal activity in the symmetrical muscles. As a consequence, the animal will move in a straight line to the source of light until some other asymmetrical disturbance once more changes the direction of motion.

What has been stated for light holds true also if light is replaced by any other form of energy. Motions caused by light or other agencies appear to the layman as expressions of will and purpose on the part of the animal, whereas in reality the animal is forced to go where carried by its legs. For the conduct of animals consists of forced movements.

The term forced movements is borrowed from brain physiology, where it designates the fact that certain animals are no longer able to move in a straight line when

---

[a] We are speaking of positively heliotropic animals exposed to only one source of light.

certain parts of the brain are injured, but are compelled to deviate constantly toward one side, which is (according to the species and the location of the injury in the brain) either the side of the injury or the opposite side. The explanation of these forced movements is that on account of the one-sided injury of the brain the tension of the symmetrical muscles is no longer the same. As a consequence, the impulses for locomotion which are equal for symmetrical muscles will cause greater contraction in certain muscles of one side of the body than in the symmetrical muscles of the other side, and the animal will no longer move in a straight line. The only difference between the forced movements induced by unequal illumination of the two eyes and by injury to the brain is that in the latter case the forced movements may last for days or throughout the whole life, while in the former case they last only as long as the illumination on the two sides of the body is unequal. If we bring about a permanent difference in illumination in the eyes, *e.g.*, by blackening one eye in certain insects, we can also bring about permanent circus motions. This shows that animal conduct may be justly designated as consisting of forced movements.

The idea that the morphological and physiological symmetry conditions in an animal are the key to the understanding of animal conduct demanded that the same principle should explain the conduct of plants, since plants also possess a symmetrical structure. The writer was able to show that sessile animals behave toward light exactly as do sessile plants; and motile animals like motile plants. The forced orientations of plants by outside sources of energy had been called tropisms; and the theory of animal conduct based on the symmetrical structure of their body was, therefore, designated as the *tropism theory of animal conduct.*

We started with symmetrical animals since in their case the analysis of conduct is comparatively simple; the results obtained in the study of these symmetrical organisms allow us also to understand the conduct of asymmetrical animals. We shall see that the principles underlying their conduct are the same as in the case of symmetrical animals, the asymmetry of the body altering only the geometrical character of the path in which the animal is compelled to move, not, however, the mechanism of conduct. While a perfectly symmetrical organism, possessed of positive heliotropism, moves in a straight line to the source of light, the path deviates from the straight line in the case of an asymmetrical organism and may in some cases, as, *e.g.,* in *Euglena,* be a spiral around the straight line as an axis. Some authors have tried to use asymmetrical organisms as a starting point for the analysis of conduct, but since it is impossible to understand the conduct of the asymmetrical organisms unless it is based upon that of the symmetrical animals, these authors have been led to anthropomorphic speculations, such as "selection of random movements" which, as far as the writer can see, cannot even be expressed in the language of the physicist.

Although the tropism theory of animal conduct was offered thirty years ago [285, 286, 287] its acceptance was delayed by various circumstances. In the first place, the majority of the older generation of biologists did not realize that not only the methods of the physicist are needed but also the physicist's general viewpoint concerning the nature of scientific explanation. In many cases the problem of animal conduct is treated in a way which corresponds more to the viewpoint of the introspective psychologist than to that of the physicist. The attempts to explain animal conduct in terms of "trial and error" or of vague "physiological states" may serve as examples. None of these attempts have led or can lead to any exact quantitative experiments in the sense of the physicist. Other biologists have still more openly adopted an anthropomorphic method of explanation. If pleasure and pain or curiosity play a rôle in human conduct, why should it be otherwise in animal conduct? The answer to this objection is that typical forced movements when produced in human beings, as, e.g., in Ménière's disease or when a galvanic current goes through the brain, are not accompanied by sensations of pleasure or pain, and there is no reason to attribute the circus movements of an animal, after lesion of the brain or when one eye is blackened, to curiosity or thrills of delight. An equally forcible answer lies in the fact that plants show the same tropisms as animals, and it seems somewhat arbitrary to assume that the bending of a plant to the window or the motion of swarmspores of algæ to the window side of a vessel are accompanied or determined by curiosity or by sensations of joy or satisfaction. And finally, since we know nothing of the sentiments and sensations of lower animals, and are still less able to measure them, there is at present no place for them in science.

The second difficulty was created by the fact that the Aristotelian viewpoint still prevails to some extent in biology, namely, that an animal moves only for a purpose, either to seek food or to seek its mate or to undertake something else connected with the preservation of the individual or the race.[b] The Aristotelians had explained the processes in the inanimate world in the same teleological way. Science began when Galileo overthrew this Aristotelian mode of thought and introduced the method of quantitative experiments which leads to mathematical laws free from the metaphysical conception of purpose. The analysis of animal conduct only becomes scientific in so far as it drops the question of purpose and reduces the reactions of animals to quantitative laws. This has been attempted by the tropism theory of animal conduct.

---

[b] This view is still held, especially among authors, who lean more or less openly to vitalism, *e.g.*, v. Uexküll, Jordan, Franz, Bauer, Buddenbrock, and others.

CHAPTER II

THE SYMMETRY RELATIONS OF THE ANIMAL BODY AS THE STARTING POINT FOR THE THEORY OF ANIMAL CONDUCT

THE starting point for a scientific and quantitative analysis of animal conduct is the symmetry relations of the animal body. The existence of these symmetry relations reduces the analysis to a comparatively simple problem.

Organisms show two forms of symmetry, radial symmetry, for which jellyfish and the stems and roots of most plants offer a well-known example, and lateral symmetry, such as exists in man and most animals. In radial symmetry the peripheral elements are distributed equally about an axis of symmetry, in the case of lateral symmetry the peripheral elements are distributed equally to the right and left of the plane of symmetry (or the median plane) by which the body is divided into a right and left half. The importance of this symmetrical structure lies in the fact that the morphological plane of symmetry is also the dynamical plane of symmetry of the organism. Symmetrical spots of the surface of an animal are chemically identical, having the same chemical constitution and also the same quantity of reacting masses. Thus the two eyes are symmetrical organs, each containing the same photochemical substances in equal quantity. In the eye itself each element is to be considered as chemically identical with the symmetrical point in the other eye. Hence, if the two eyes are illuminated equally, the photochemical reaction products produced in the same time will be equal in both eyes. What is true for the eyes is true for all symmetrical elements of the surface of an animal.

The median plane is also the plane of symmetry for the muscles and the muscular activity of the body. Symmetrical muscles possess under equal conditions equal tension and symmetrical muscles are antagonistic to each other in regard to moving the body to the right or left.

We say that impulses go from the central nervous system to the muscles; and from the surface of the body to the central nervous system. According to our present knowledge that which is called a nervous impulse seems to consist of a wave of chemical reaction traveling along a nerve fiber. The central nervous system is also symmetrical and, moreover, we may conceive a projection of the elements of the surface of the body upon the ganglion cells and from here to the muscular system of the body. The complications in this system of projections constitute the difficulties in our understanding of the structure of the brain. This idea of a projection of the sense organs or the surface of the body upon the brain will explain why the morphological plane of symmetry of an organism is also its plane of symmetry in a dynamical sense. When symmetrical elements of the eyes are struck by light of the same wave length and intensity, the velocity of photochemical reactions will be the same in both eyes. Symmetrical spots of the retina are connected with symmetrical elements in the brain and these in turn are connected with symmetrical muscles. As a consequence of the equal photochemical reactions in the symmetrical spots of the retina equal changes are produced in the symmetrical brain cells with which they are connected, and equal

SYMMETRY RELATIONS

changes in tension will be produced in the symmetrical muscles on both sides of the body with which the active brain elements are connected.[a] On account of the symmetrical character of all the changes no deviation from the original direction of motion will occur. If, however, one eye is illuminated more than the other eye, the influence upon the tension of symmetrical muscles will no longer be the same and the animal will be forced to deviate from the original direction of motion.

We have thus far considered only the relation between right and left. Aside from this symmetry relation we have polarity relations, between apex or head and base or tail end. Just as we found that the morphological plane of symmetry is also a dynamical plane of symmetry, we find that with the morphological polarity head-tail is connected a dynamic polarity of motion, namely, forward and backward. This will become clear in the next chapter on forced movements.

Physiologists have long been in the habit of studying not the reactions of the whole organism but the reactions of isolated segments (the so-called reflexes). While it may seem justifiable to construct the reactions of the

---

[a] Physiologists assume that stimulations are constantly sent from the brain to the muscles and that this maintains their tension. v. Uexküll uses the term that "tonus" is sent out to the muscle and that the brain is a reservoir of "tonus" as if the latter were a liquid. The writer wonders whether it might not be wiser to substitute for such metaphors hypotheses in terms of chemical mass action. Constant illumination causes a stationary process in photosensitive elements of our eye, in which the mass of the reaction product is determined by the Bunsen-Roscoe law. We assume, moreover, that in proportion to this photochemical mass action corresponding chemical reactions take place in the brain elements with which the eyes are connected; and that as a consequence corresponding chemical reactions take place in the muscles by which the tension of the latter is determined. These processes in the muscles may possibly consist in the establishment of a definite hydrogen ion concentration. Such hypotheses have the advantage over the "stimulation" hypothesis that they can be tested by physicochemical measurements.

organism as a whole from the individual reflexes, such an attempt is in reality doomed to failure, since reactions produced in an isolated element cannot be counted upon to occur when the same element is part of the whole, on account of the mutual inhibitions which the different parts of the organism produce upon each other when in organic connection[b]; and it is, therefore, impossible to express the conduct of a whole animal as the algebraic sum of the reflexes of its isolated segments.

E. P. Lyon[320] has shown that if the tail in a normal shark be bent to one side (without changing the position of the head) the eyes of the animal move as promptly as compass needles in association with the bent tail around the same axis in which the bending occurs, but in an opposite sense. On the convex side of the animal, the white of the eye is more visible in front, on the concave side it is more visible behind; hence the former has moved backward, the latter forward. This was observed not only in the normal fish but also when the optic and auditory nerves were cut. The central nervous system acts as one unit. R. Magnus[332] and his fellow-workers have shown that an alteration in the position of the head of a dog inevitably alters the tone of the muscles of the legs.[c] These and other associations and mutual inhibitions make possible that simplification which allows us to treat the

---

[b] When the stem of a plant (*e.g.*, *Bryophyllum*) is cut into as many pieces as there are nodes, each node will under the proper conditions give rise to one or two shoots. If we leave them in connection, only the buds at the apical end will grow out, the rest of the buds remaining dormant. The whole stem acts as though it consisted of only the bud situated at the apex.

[c] The problem of coördination will form the subject of another volume in this series by Professor A. R. Moore, and for this reason a fuller discussion of work on coördination, such as that by Sherrington and by v. Uexküll, may be reserved for Professor Moore's volume.

organism as a whole as a mere symmetry machine, a simplification which forms the foundation of the tropism theory of animal conduct.

It would, therefore, be a misconception to speak of tropisms as of reflexes, since tropisms are reactions of the organism as a whole, while reflexes are reactions of isolated segments. Reflexes and tropisms agree, however, in one respect, inasmuch as both are obviously of a purely physico-chemical character.

CHAPTER III

FORCED MOVEMENTS

WHEN we destroy or injure the brain on one side we paralyze or weaken the muscles connected with this side. As a consequence the morphological plane of symmetry ceases to be the dynamical plane of symmetry and the animal has a tendency to move in circles instead of in a straight line. Suppose a fish swimming forward by motions of its tail fin. Normally the stroke occurs with equal energy to the right and to the left, and the rudder action of the tail is equal in both directions, but after the lesion of one side of the brain the stroke and the rudder action cease to be the same in both directions, it is weakened in one direction. Hence the animal instead of swimming in a straight line is forced to deviate continually toward one side from the straight line of locomotion. We speak in such a case of a forced motion.

When we destroy the ventral portion of the left optic lobe in a shark (*Scyllium canicula*), the fish no longer swims in straight lines but in circles to the right (when the right optic lobe is destroyed it swims in circles to the left). After the destruction of the left optic lobe, the muscles on the left side of the tail are weakened or semiparalyzed, and they no longer produce the same rudder action as the muscles on the right side. Hence the impulses (or nerve processes) which flow in equal intensity to the muscles on both sides will no longer produce equally energetic rudder action of the tail to the right and to the left, but the muscles turning the tail to the right will contract more powerfully than those turning it in the opposite direction. The outcome of this greater rudder action of the tail when moving to the right is that the fish instead of swimming in a straight line moves in a circle to the right.[290]

It is often the case that the body of such a fish even when quiet is no longer straight but bent in a circle, the left side forming the convex side; and when such a fish dies and rigor mortis sets in it may become stiff in this position. These latter observations prove that the circus movements to the right are due to the lowering of the tension of the lateral muscles of the body on the left side of the fish. This is the fundamental fact for the theory of forced movements—namely, that a lesion in one side of the brain lessens the tension of the muscles on one side of the body; as a consequence the motions of the animal become difficult or impossible in one direction and become easy in the opposite direction.

In many cases the motions of an animal depend upon a coöperative activity of two sets of appendages, *e.g.*, the pectoral fins of a fish or the legs of an animal. Such coöperative or associated action is determined by the fact that the same nerve center supplies antagonistic muscles of the two organs (*e.g.*, the lateral fins). Thus the same nerve impulse causes both our eyes to move simultaneously to the right or to the left. When we look to the right, the same impulse which causes the contraction of the rectus externus muscle in the right eye causes a contraction of the rectus internus muscle in the left eye. These two muscles then are associated.

In a fish like the shark the position and innervation of the eyes differ from that of the human being. In the shark the eyes are not in front but on the side, and the muscles which lift the eye on one side are associated with those which lower it on the other side of the head. A similar association exists in regard to the pectoral fins, the muscles which lift the right pectoral fin are associated with those which lower the left one, and vice versa. When a normal shark swims the two pectoral fins work equally and the fish swims without rolling over to the right or to the left.

If we destroy in a shark the left side of the medulla oblongata forced changes in the position of the two eyes and the two pectoral fins will follow.[290] (There are in addition correlated changes in the other fins and the rest of the body which we will omit in order to simplify the presentation of the subject.) When a shark, whose left medulla is cut, is kept in a horizontal position, its left eye looks down and the right eye looks up. This change of position of both eyes indicates that the relative tension between the muscles of the eyes has changed. In the left eye the tension of the lowering muscles predominates over that of their antagonists, in the right eye the reverse is the case. The pectoral fins likewise show associated changes of position. The left fin is raised up dorsally, the right is bent down ventrally. Since we know that the destruction of the central nervous system causes a paralysis of muscles and not the reverse we must conclude that the destruction of the left side of the medulla in a shark causes a weakening or partial paralysis of the muscles which lower the left fin and of those which raise the right fin. Hence the muscles which press down on the water will press harder in the right than in the left fin. When such an animal swims rapidly, it will come under the influence of a couple of forces which must produce a rolling movement around the longitudinàl axis of its body toward the left. These rolling motions are another well-known type of forced movements. When such an animal swims slowly, it will roll more than a normal fish, but it will not roll completely around its longitudinal axis. These are the same motions which were observed in dogs by Magendie and Flourens [155] after an operation in the medulla or pons. We can state, therefore, that the rolling motions are caused by the weakening of one group of (associated) muscles while their antagonists are not weakened.

It is of interest to consider the nature of forced movements after injury of the cerebral hemispheres in a dog. When in a dog one of the cerebral hemispheres is injured the animal immediately after the operation no longer moves in a perfectly straight line, but deviates from the straight line toward that side where the brain is injured.[178] When the left hemisphere is injured circus motions toward the left ensue. If one offers a dog which was operated in the left cerebral hemisphere a piece of meat, removing it as fast as the dog approaches, the dog will move at first a certain distance in a straight line; it will then suddenly turn to the left and describe a complete circle, moving afterward for a little while in a straight line toward the meat and turning again through an angle of 360° to the left, and so on.[284] The explanation is the same as for the foregoing cases. The lesion of the left cerebral hemisphere caused a weakening or partial paralysis of the muscles which turn the body to the right. Hence the animal will, when following the meat, deviate to the left, and this causes a displacement of the retina image in the same direction and an apparent motion of the object to the right. We shall see in a later chapter on the orienting effect of moving retina images that this deviation of the retina image to the left causes a forced motion of the animal to the right which compensates its tendency to deviate to the left due to the effect of the brain lesion. Hence the animal approaches the meat in an approximately straight line. But it does so with difficulty and sooner or later tiring of this effort it moves in the usual automatic way, whereby equal impulses reach the muscles on both sides. This results in a complete circus movement to the left on account of the weakening (caused by the operation) of muscles which turn the body to the right. The retina image of the meat again induces a straight motion and the whole process described is repeated. When the injury to the brain was less severe the animal may follow the meat for long distances without turning in a circle.

When such a dog is offered simultaneously two pieces of meat, one in front of the left, the other in front of the right eye, it invariably moves toward the one on the left side. The equal flow of impulses caused by the symmetrically located pieces of meat results in a stronger contraction in the muscles on the left than on the right side of the body, since as a consequence of the lesion the tension of the former muscles is greater than that of the latter. When two pieces of meat are simultaneously offered to the dog, but both pieces are in front of the left eye, the dog tries to get the piece nearest to its mouth, but its effort carries it a little too far to the left and then it takes the other piece of meat which is situated farther to the left.[284]

Some time after the operation these disturbances may become less and may ultimately disappear. If now the dog is operated on the other, *e.g.,* the right hemisphere, circus motions to the right appear.

We do not wish to exhaust the chapter on forced movements but may perhaps for the sake of completeness point out the following facts. We have seen that if one cerebral hemisphere is injured the dog shows a tendency to circus movements to the operated side. When both hemispheres are injured, *e.g.,* both occipital lobes are removed, the dog can hardly be induced to move forward and it is impossible to cause it to go downstairs, while it is willing to go upstairs. Its front legs are extended and its head is raised high, giving the impression as if such a dog had a tendency to move backward rather than forward or that the forward movement was difficult. If the two anterior halves of the cerebral hemispheres are removed the reverse happens. The animal runs incessantly as if driven by a mad impulse; its head is bent down and it is in every respect the converse of the animal operated in the occipital lobes. These two types of forced movements correspond to the morphological polarity tail-head. This corresponds to the idea of a projection of the surface elements upon the brain either directly or by crossing.

These three types of forced movements: the circus motions, the tendency to go backward, and the irresistible tendency to move forward will appear in the form of the tropistic reactions to be described in this volume.

Since we shall deal in this volume chiefly with invertebrates, it may be of importance to show that forced movements can also be produced in this group of animals by lesion of one side of the cerebral ganglion, and that these forced movements depend also upon the fact that as a consequence of the operation the tension of symmetrical muscles (which is equal under normal condi-

30   TROPISMS

tions) becomes unequal. Fig. 1, *B*, gives the change in position of the body and of the legs in the larva of a dragon fly (*Æschna*) after the left half of the cerebral ganglion has been destroyed (Matula [541]). Such an animal moves in a circle to the right. The longitudinal muscles connecting the segments of the body are under higher tension on the right side of the body than on the left and the body

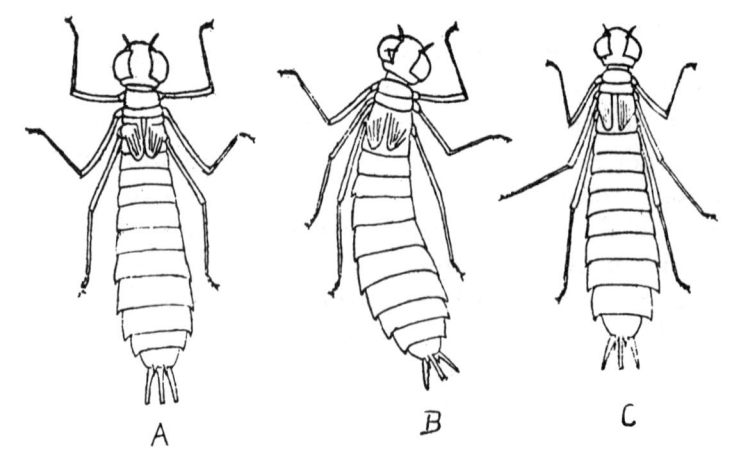

Fig. 1.—*B*, forced position of larva of the dragon fly (*Æschna*) whose left cerebral ganglion is destroyed. The body is convex on the left side, due to a relaxation of the muscles connecting the segments on the left side. The position of the legs is such that the animal can only move in circles to the right. This asymmetry disappears again when both ganglia are destroyed, *C*. *A*, normal animal. (After Matula.)

is bent with its convex side to the left. The normally symmetrical position of the legs (Fig. 1, *A*) is now changed in such a way (Fig. 1, *B*) that the animal is no longer able to move in a straight line, but is forced to move in circles to its right. We shall see later that similar changes in the position of the legs are produced in a positively heliotropic insect when the left eye is blackened and in a negatively heliotropic insect when the right eye is blackened. Circus motions after destruction of one cephalic ganglion in an insect are a general occurrence and have been known for a long time.

The importance of these forced movements caused by lesion of the brain for the explanation of the conduct of normal animals lies in the fact that the latter is essentially a series of forced movements. The main difference between the forced movements after brain lesion and the conduct of a normal animal lies in the fact that the former are more or less permanent; while in the normal animal conduct the changes in the relative tone of symmetrical muscles leading to a temporary forced movement are caused by a difference in the velocity of chemical reactions in symmetrical sense organs or other elements of the surface.

CHAPTER IV

GALVANOTROPISM

WHEN we send a galvanic current lengthwise through a nerve, at the region near the anode the irritability of the nerve is diminished, while it is increased near the cathode. The condition of decreased irritability near the anode is called anelectrotonus and the increased irritability near the cathode is called catelectrotonus. When a current is sent through an animal, those nerve elements which lie in the direction of the current will have an anelectrotonic and a catelectrotonic region; while the nerves through which the current goes at or nearly at right angles are not affected. Ganglia or nerve tracts in the anelectrotonic condition will, therefore, act as if they were temporarily injured, and hence we need not be surprised to find that the galvanic current causes forced movements which last as long as the current lasts, and which cease with the current.

Hermann reported in 1885 [204] that when a current is sent through a trough containing tadpoles of a frog, the tadpoles orient themselves in the direction of the current curves putting their heads to the anode.[a] Blasius and Schweizer [523] found soon afterwards that a large number of animals when put into a trough with water through which a galvanic current passes have a tendency to go to the anode. The explanation given by Hermann and by Blasius and Schweizer is not correct. They

---
[a] The writer has never been able to repeat this observation.

GALVANOTROPISM

assumed that the current, acting upon the central nervous system, causes sensations of pain when it goes in the direction from tail to head in the animal; while it has a soothing or hypnotizing effect when it goes in the opposite direction, namely from head to the tail. In the latter case the head is directed toward the anode. The authors assume that the animals choose the position with least pain, *i.e.*, with their heads to the anode. This assumption is wrong, since we know that when a galvanic current is sent through the head of a human being automatic motions comparable to those observed in animals occur which are not voluntary and which are unaccompanied by any pain sensation. Thus when a galvanic current is sent laterally through the head, the person falls toward the anode side but has no feeling of pain. Mach noticed the same effect of falling to the side of the anode when a galvanic current was sent sidewise through fishes.[330] These galvanotropic motions are in reality forced movements, and this has been proved by direct observations. It was shown by Loeb and Maxwell[307] in experiments on crustaceans and by Loeb and Garrey[306] on salamanders that when we send a galvanic current through animals which go to the anode, changes in the position of the legs occur comparable to the changes in the position of fins and eyes mentioned in the previous chapter, and that these changes are of such a character as to make it easy for the animal to move in the direction of the anode and difficult if not impossible to move in the opposite direction.

In all these experiments it is of importance to choose the proper density of the current. For the experiments on the shrimp (*Palæmonetes*)[307] the animals were put into a square trough, two opposite sides of which were formed of platinum electrodes. The cross section of the *fresh* water in the trough was 1,400 mm.² and the intensity of the current about 1 milliampere or a little less. We found it advisable to increase the intensity very gradually by increasing slowly the resistance of a rheostat in a short circuit until the phenomenon of galvanotropism appeared most strikingly. When the current is too strong or too weak the phenomena are no longer clear. The common shrimp (*Palæmonetes*) is a marine crustacean which

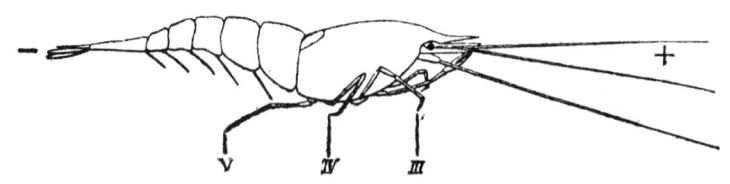

FIG. 2.—Forced position of shrimp (*Palæmonetes*) when galvanic current goes from head to tail. Tension of extensor muscles of tail fin prevails over that of flexors. Animal can swim forward (to anode), but not backward. (After Loeb and Maxwell.)

lives also in brackish water and which can stand fresh water long enough for the purpose of these experiments. The animal can swim forward as well as backward; in forward swimming the extensor muscles of its tail fin work more strongly than the flexors (Fig. 2); in swimming backward the flexors work energetically (Fig. 3) and thus produce a powerful stroke forward, while the extensors contract with less energy. When we put a *Palæmonetes* in a trough through which a current goes and if we put the animal with its head toward the anode the tail is stretched out (Fig. 2). This means that the tension of the extensor muscles prevails over that of the flexors and since the forward swimming is due to the stroke of

GALVANOTROPISM

the extensors, and since it is antagonized by the tension of the flexors, the animal can swim forward but not backward, or only with difficulty; if we put the animal with its head toward the cathode the tail is bent ventrally (Fig. 3), which means that the tension of the flexors is stronger than that of the extensors. As a consequence the animal can swim backward but not forward, or only

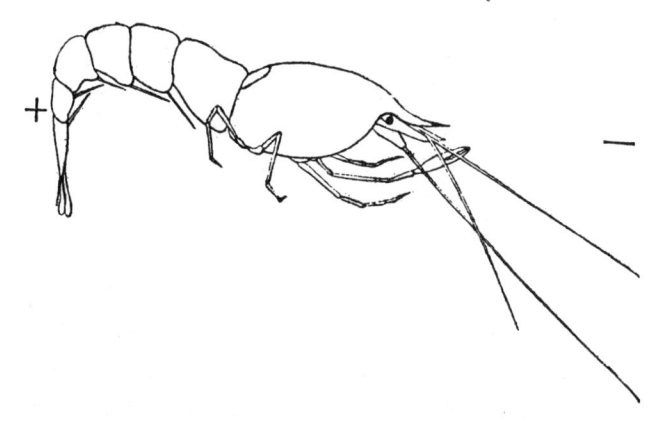

Fig. 3.—Forced position of shrimp when positive current goes from tail to head. Tension of flexors of tail fin prevails over that of extensors. Animal can swim backward (to anode), but not forward. (After Loeb and Maxwell.)

with difficulty. In both cases the result will be a swimming of the animal to the anode, in the former case by swimming forward in the latter by swimming backward.

We can further show that the tension of the muscles of the legs of *Palæmonetes* is always altered in such a sense by the galvanic current that motion toward the anode is facilitated, while that toward the cathode is rendered difficult or impossible. The animal uses the third, fourth, and fifth pair of legs for its locomotion (Fig. 2). The third pair pulls in the forward movement and the fifth pair pushes. The fourth pair acts like the fifth and requires no special discussion. If a current be sent through the animal longitudinally from head to tail and the intensity be increased gradually, a change soon takes place in the position of the legs. In the third pair the tension of the flexors predominates (Fig. 2), in the fifth the tension of the extensors. The animal can thus move easily by pulling of the third and by pushing of the fifth pair of legs, that is to say, the current changes the tension of the muscles in such a way that the forward motion is facilitated, while the backward motion is rendered difficult. Hence it can easily go toward the anode but only with difficulty toward the cathode. If a current be sent through the animal in the opposite direction, namely from tail to head, the third pair of legs is extended, the fifth pair bent (Fig. 3); *i.e.*, the third pair can push, the fifth pair can pull backward. The animal can thus go backward with ease but forward only with difficulty. This again will lead to a gathering of such animals at the anode, this time, however, by walking backward.

The phenomena thus far described recall the forced movements mentioned in the third chapter, where certain injuries of the brain accelerate forward motion while other lesions in the opposite parts of the brain make forward motion difficult if not impossible.

Palæmonetes can also walk sidewise. This movement is produced by the pulling of the legs on the side toward which the animal is moving (contraction of the flexors), while the legs of the other side push (contraction of extensors). If a current be sent transversely, say from left to right, through the animal, the legs of the left side assume the flexor position, those of the right side the

GALVANOTROPISM

extensor position (Fig. 4). The transverse current thus makes it easy for the animal to move toward the left—the anode—and prevents it from moving toward the right—the cathode. If a galvanic current flows transversely

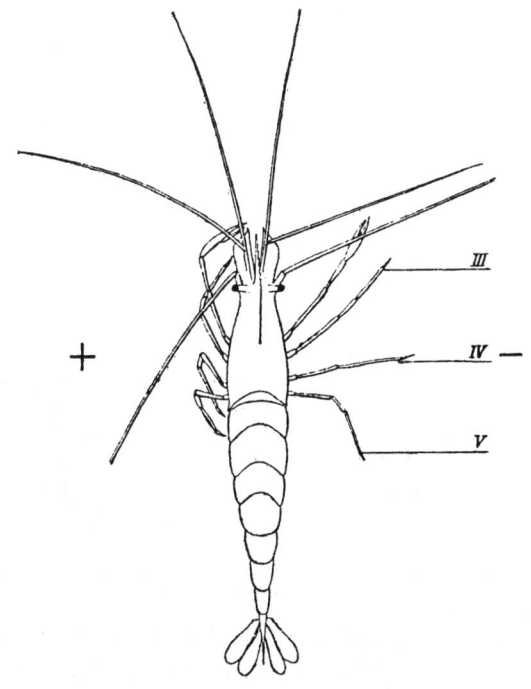

Fig. 4.—Position of legs of shrimp when current goes sidewise through the animal, from left to right. In the legs on the left side the tension of the flexors, in those of the right side the tension of the extensors predominates. The animal can easily go to the left (anode), but not to the right. (After Loeb and Maxwell.)

through the animal, it creates the analogue of the circus motions produced by injury of one side of the brain. Figs. 5 and 6 show that the current produces similar effects in the crayfish as those produced in the shrimp (Figs. 2 and 3).

It is not difficult to suggest by aid of a diagram the arrangement of the elements in the central nervous system required to bring about the phenomena of galvanotropism just described for *Palæmonetes*. We take it for granted that the regular phenomena of anelectrotonus and catelectrotonus of motor nerve elements suffice for the explanation of these phenomena. We assume that if the cell body of a neuron is in the state of catelectrotonus

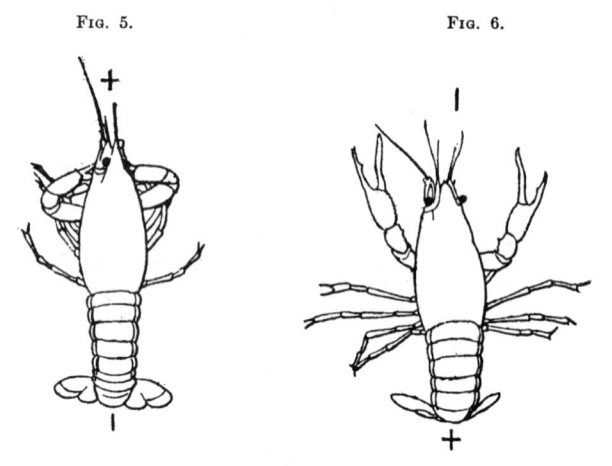

FIG. 5. FIG. 6.

FIGS. 5 and 6.—Show the same effects of current on the common crayfish as those on *Palæmonetes* in Figs. 2 and 3.

its activity is increased, when it is in anelectrotonic condition activity is diminished. Neurons having the same orientation will always be affected in the same sense by the current.

Fig. 7 is a diagram of the arrangement of neurons giving rise to the bending of the legs on the side of the anode and to the extension of the legs on the side of the cathode when the current goes sidewise through the animal. This diagram assumes that the nerves innervating the extensors come from the opposite side of the central

GALVANOTROPISM

nervous system, while those innervating the flexors are on the same side. This diagram corresponds to reality, according to the histological work of Allen. When the current goes from right to left through the crustacean the cell bodies of the neurons on the right side are in catelectrotonus, those on the left side in anelectrotonus. The former are, therefore, in a state of increased "irritability," the latter in a state of diminished "irritability." Hence the flexors of the right leg are contracted and the extensors relaxed, while the flexors of the left leg are relaxed and the extensors contracted.

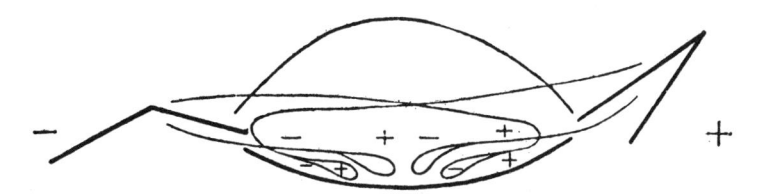

Fig. 7.—Diagram indicating the orientation of the neurons for flexor and extensor muscles of the right and left legs to explain changes of position of legs under influence of galvanic current. (After Loeb and Maxwell.)

Another crustacean *Gelasimus*[307] shows the same effect of the current when it goes sidewise through its body. When the thoracic ganglion from which the nerves of the legs originate is cut longitudinally in the middle, all the legs assume permanently a bent position, confirming our assumption that the extensor nerves cross over while the flexors originate from the same side of the ganglion on which their muscles are. It, therefore, looks as if our diagram were the expression of the actual condition.

In the same way we can explain the results of a galvanic current when it goes through the animal lengthwise. We only need to assume that the cell bodies which send their fibers to the flexors of the third pair of legs have the same orientation as the cell bodies which send their fibers to the extensors of the fifth pair of legs (Fig. 8). Hence when the positive current goes from head to tail through the animal (Fig. 8), the flexors of the third pair of legs and the extensors of the fifth pair must be thrown into greater activity, since the cell bodies of both these nerves are in a condition of catelectrotonus, *i.e.*, increased activity.

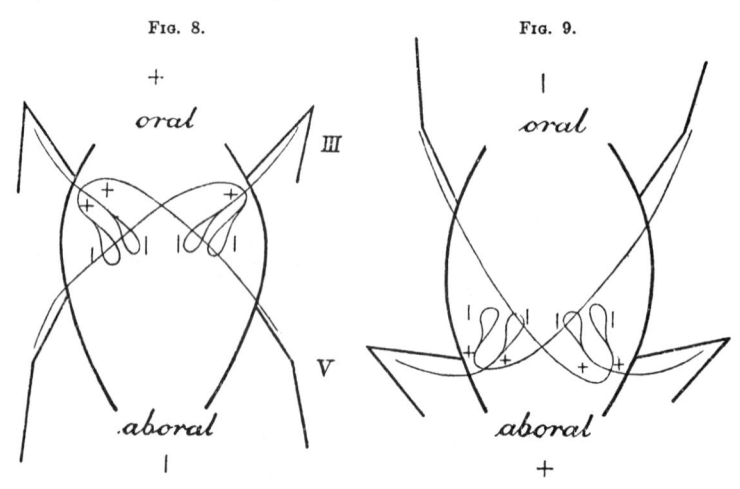

FIG. 8.   FIG. 9.

FIGS. 8 and 9.—Diagram indicating orientation of neurons for flexor and extensor muscles of third and fifth pairs of legs to explain galvanotropic reaction. (After Loeb and Maxwell.)

When the current goes from tail to head the cell bodies of the extensors of the third and of the flexors of the fifth pair of legs are in catelectrotonus. This possibility is expressed in the diagram Fig. 9.

In this way the theory of the galvanotropic reaction of those animals which go to the anode seems complete.

What has been demonstrated for *Palæmonetes* holds not only for many crustaceans but for vertebrates also. Loeb and Garrey[306] have shown that when a current

GALVANOTROPISM

is sent through a trough containing larvæ of a salamander (*Amblystoma*) the legs and head of the larvæ assume definite positions depending upon the direction of the current. When the current goes from head to tail the legs are pushed backward and the head is bent (Fig. 10);

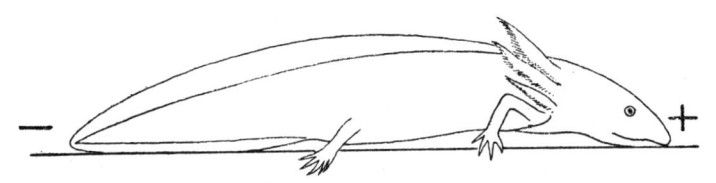

Fig. 10.—Forced position of *Amblystoma* larva under influence of galvanic current going through animal from head to tail. Head down, body convex on dorsal side. Legs backward. (After Loeb and Garrey.)

when the current goes from tail to head the opposite position is observed (Fig. 11). The analogy with the observations on *Palæmonetes* is obvious.

Galvanotropic reactions are found throughout the whole animal kingdom and the following observations made by Bancroft on a jellyfish (*Polyorchis penicillata*)

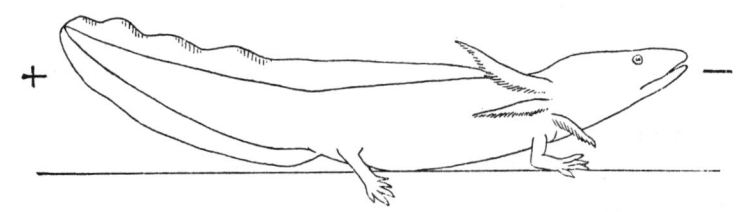

Fig. 11.—Forced position of *Amblystoma* larva when current goes from tail to head. Head raised, legs pushed forward, tail raised. (After Loeb and Garrey.)

are of especial interest.[16] Strips containing tentacles and the manubrium were cut out from the animal and put into a trough through which a current flowed of 25 to 0.200 m. a. for 1 square mm. of the cross section of the liquid (dilute sea water) in the trough.

If a meridional strip passing from the edge on one side through the center of the bell to the other edge be prepared and the current passed through transversely, tentacles and manubrium turn and point toward the cathode (Fig. 12). A reversal of the current initiates a turning of these organs in the opposite direction, which is usually completed in a few seconds. This can be repeated many times and the tentacles continue to respond after hours of activity. The manubrium, however, tires sooner and fails to respond. If the strip is placed with its subumbrella surface upward and extended in a straight line parallel to the current lines, the making of the current causes the tentacles at the anode end to turn through an angle of 180° and point toward the cathode. The tentacles at the cathode end become more crowded together, reminding one of the tip of a moistened paint brush, and also point more directly toward the cathode (Fig. 13). The experiment may be varied in still other ways by cutting smaller or larger pieces from the edge of the swimming bell, but the response is always the same. The tentacles, wherever possible, and to a less extent the manubrium, bend so as to point toward the cathode. The response depends in no way upon the connection of these organs with the swimming bell, muscles, or nerve ring, for it is obtained equally well with isolated tentacles and pieces of tentacles. Isolated tentacles when placed transversely to

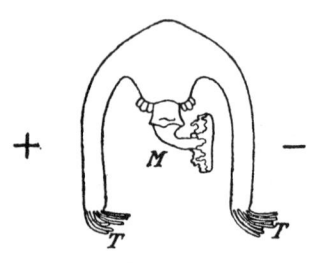

Fig. 12.—Tentacles *T* and manubrium *M* of a jellyfish (*Polyorchis*) under influence of galvanic current are turned to the negative pole. (After Bancroft.)

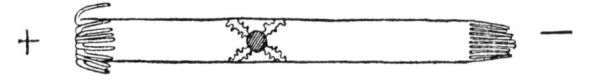

Fig. 13.—Strip of jellyfish showing that under the influence of galvanic current tentacles on both ends point towards cathode. (After Bancroft.)

the current lines curve so as to assume a more or less complete U-shape, with their concave side toward the cathode. When placed parallel to the current, the tentacles do not curve.[16]

The latter observation shows the fact that the whole reaction is due merely to an increase in the tension of the muscles on the cathode side of the organ.

Phenomena of galvanotropism can be observed also in infusorians. Thus Verworn[493] observed that when

GALVANOTROPISM

a current goes through a trough containing *Paramæcia* the animals will all move toward the cathode. The mechanism of the reaction was discovered by Ludloff.[317] The locomotion of *Paramæcium* is brought about by cilia. As a rule the cilia are directed backward (Fig. 14), and in their normal movement they strike powerfully backward and are retracted with less energy to their normal position. Since their powerful stroke is backward the animal is pushed forward in the water. Ludloff and Bancroft [17, 18] show that if a *Paramæcium* is struck sidewise by the current, the position of the cilia on the cathode side is reversed inasmuch as they are now turned forward. On the anode side they continue to be directed backward (Fig. 15, *a*). Instead of striking symmetrically on both sides of the animal, the cilia on the cathode side strike forward powerfully while those on the anode side strike powerfully backward. The animal is thus under the influence of a couple of forces which turn its oral pole toward the cathode side. As soon as it is in this condition the symmetrical cilia are struck at the same angle by the parallel current lines and they must assume a symmetrical position which is as in Fig. 15, *b*, namely the cilia are pointed forward toward the cathode at the oral end, and backward toward the anode at the aboral end. As long as the current is not too strong, the oral region, where the cilia are pointing forward, is rather small and therefore the action of those cilia prevails which are in the majority and which are pointed backward. As a result the organism moves slowly forward to the cathode.

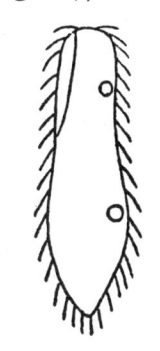

Fig. 14.—*Paramæcium* under normal conditions. Cilia all pointing toward aboral pole.

TROPISMS

A similar mechanism of galvanotropic conduct exists in *Volvox* a spherical, unicellular organism which is surrounded by cilia on its whole surface. A definite pole of the organism is always foremost in all locomotions. This organism usually swims to the anode when in a galvanic field. Bancroft made the action of the cilia of *Volvox* visible with the aid of india ink and was able to show that the current made the cilia on the anode side stop, while those on the cathode side continue to beat.[20]

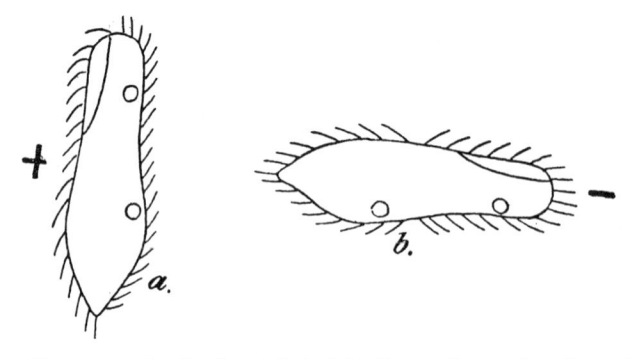

Fig. 15.—*a*, current going from left to right through *Paramœcium*, the position of cilia on the cathode side is now reversed, their free ends pointing forward. The animal when swimming is automatically turned with its oral end toward the cathode. *b*, current going through *Paramœcium* from aboral to oral end. Cilia symmetrical on both sides but pointing forward at oral end and backward at aboral end.

Since the backward stroke is always the effective one the organism is thus carried automatically toward the anode.

Terry[478] found that *Volvox* can be made to move toward the anode as well as toward the cathode. It moves to the anode after having been kept in the dark for two or three days, while after exposure to light it swims to the cathode. *Volvox* contains chlorophyll and the change in the sense of reaction is therefore connected with the formation of a product of chlorophyll activity. Bancroft found that when *Volvox* was made cathodic by exposure to sunlight, the cilia stop on the cathode side.

GALVANOTROPISM

While the locomotor mechanism of unicellular organisms, like *Paramæcia* and *Volvox,* is as simple as that of higher organisms, the locomotion of microörganisms possessing only one flagellum, like *Euglena,* is more complicated. It was generally assumed that the flagellum acted like a single oar and that it was directed forward, but this is not correct. It is shaped like a U and its free end is directed backward; and Bancroft has emphasized that it acts by the formation of a loop which moves like a wave from the base of the flagellum to its free tip. The same author discovered that *Euglena* are galvanotropic when raised in acid media. On account of the asymmetry of their locomotor apparatus they are compelled to swim in a spiral, in most cases to the cathode, exceptionally to the anode. Bancroft showed that the orientation of these organisms by the galvanic current is identical with that by light.[21]

All the phenomena of galvanotropism are, therefore, reduced to changes in the tension of associated muscles or contractile elements, as a consequence of which the motion of the organism toward one pole is facilitated, while the motion toward the opposite pole is rendered difficult. Galvanotropism is, therefore, a form of forced motions produced by the galvanic current instead of by injury to the brain.

There remains then the question of how a galvanic current can bring about those changes which result in the anelectrotonic and catelectrotonic condition mentioned at the beginning. Currents can pass through tissues only in the form of ions whose progress is blocked by membranes which are more permeable for certain salts than for others. Those salts which go through the membrane carry the current through the tissue elements, those which do not go through will increase in concentration at the surface of the membrane. It is the latter which cause the electrotonic effects; according to Loeb and Budgett [304] by secondary chemical reactions at the boundary. Nernst has pointed out that a stationary condition must arise at the surface of the membrane due to the fact that the increase in concentration of ions by the electric current gives rise to a current of diffusion of salt in the opposite direction away from the membrane. "The average change of concentration at the membrane depends, therefore, upon the antagonistic effects of the current and of the diffusion." [524] This must be kept in mind since otherwise the effect of the constant current should increase constantly with its duration, which is not the case, on account of the establishment of a condition of equilibrium between the increase of the concentration of ions at the boundary with the duration of the current and the diminution of this concentration by the diffusion of the ions in the opposite direction due to osmotic pressure.

CHAPTER V

HELIOTROPISM

THE INFLUENCE OF ONE SOURCE OF LIGHT

1. GENERAL FACTS

THE fact that certain animals go to the light had, of course, been known for hundreds of years, but this was explained in an anthropomorphic way. Thus Lubbock, and Graber,[180] had taken it for granted that certain animals went to the light or away from it on account of fondness for either light or darkness, and their experiments were calculated to demonstrate this fondness. Animals were distributed in a box, one-half of which was covered with common window glass, the other with an opaque body or with colored glass, and after a while the number of animals in each half was counted. When the majority of animals were found in the dark part the animals were believed to have a preference for darkness, when in the light part they were believed to be fond of light. The same method was used to decide whether animals preferred blue to red or vice versa. The writer attacked the problem from the physical viewpoint, assuming that the animals are "fond" neither of light nor of "darkness," but that they are oriented by the light in a similar way as plants are; being compelled to bend or—as in the case of motile algæ—move automatically either to a source of light or away from it.[285, 287]

In the case of unequal illumination of the two eyes the tension of the symmetrical muscles in an animal becomes unequal. In this condition the equal impulses of locomotion will result in an unequal contraction of the muscles on both sides of the animal. As a consequence the animal will turn automatically until its plane of symmetry is in the direction of the rays of light. As soon as this happens the illumination of both eyes and the tension of symmetrical muscles become equal again and the animal will now move in a straight line—either to or from the source of light. What appeared to the older authors as the expression of fondness for light or for darkness was according to the writer's theory the expression of an influence of light upon the relative tension of symmetrical muscles.

Animals which are compelled to turn and move to the source of light the writer called positively heliotropic, those which are compelled to turn and move in the opposite direction he called negatively heliotropic. The designation heliotropism (or phototropism) was chosen to indicate that these reactions of animals are of the same nature as the turning of plants to the light; and the writer was indeed able to show that sessile animals bend to the light as do plants which are raised near a window;[288] while motile animals move to (or from) a source of light as do the motile swarmspores of algæ or motile algæ themselves.

We will first discuss positively heliotropic motile animals. The positively heliotropic caterpillars of *Porthesia chrysorrhœa*[288] or the winged plant lice of *Cineraria*[288] or the newly hatched larvæ of the barnacle[183] were used by the writer in his earliest experiments and they may serve as examples. The larvæ of *Porthesia* must be used after hibernation before they have taken food. When about 50 or 100 of such larvæ are put into a test tube and the latter is placed at right angles against a window, all the animals begin to move to the window in as straight a line as the imperfections of their locomotion and collisions permit. As soon as they reach the window side of the test tube they remain there permanently, unless the test tube is turned around. If we turn the test tube around an angle of 180° the animals go at once to the window again. They react in this way whether the source of light is sunlight, diffused daylight, or lamp light; and this can be repeated indefinitely. The animals are slaves of the light. These experiments are typical for positively heliotropic motile animals.

When the animals have reached the window end of the test tubes they remain there, since the light prevents them from going back. But in staying there they may assume any kind of orientation, thus proving that the light orients them only as long as they are in motion. The light affects the tension of the muscles and we shall see later that when the animals are not moving, the change in the tension of the muscles manifests itself by changes in the position of the legs, which is noticeable in organisms with comparatively large appendages.

That these animals do not go to the light because they prefer light to darkness but because the light orients them is proved by the fact that they will go from light into the shade if by so doing they remain oriented with their heads toward the source of light.[287] Let direct sunlight S fall upon a table through the upper half of a window (W, Fig. 16), the diffused daylight D through the lower half. A test tube ac is placed on the table in such a way that its long axis is at right angles with the plane of the window; and one-half ab is in the direct sunlight, the other half in the shade. If at the beginning of the experiment the positively heliotropic animals are in the direct sunlight at *a*, they promptly move toward the window, gathering at the window end *c* of the tube, although by so doing they go from the sunshine into the shade. This experiment shows also that it is not the intensity

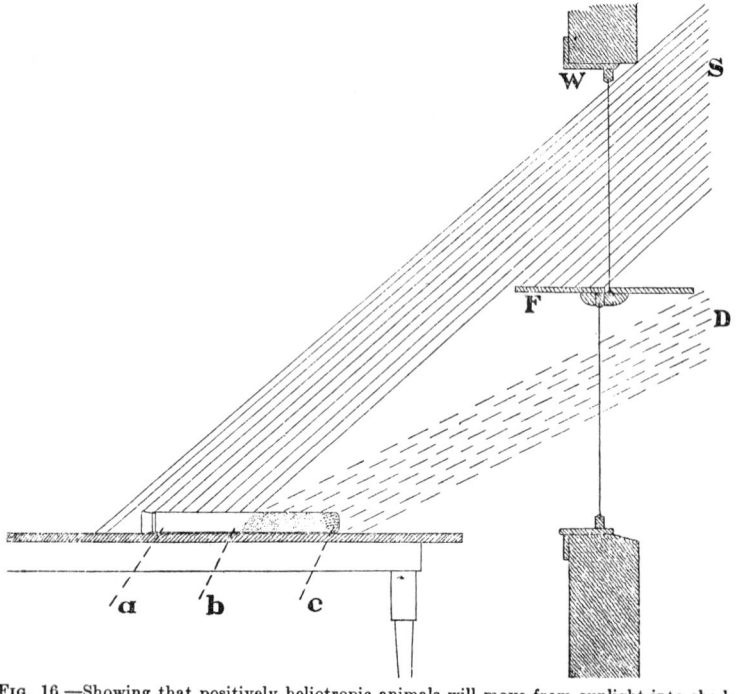

Fig. 16.—Showing that positively heliotropic animals will move from sunlight into shade if in so doing the illumination of the two eyes remains the same.

gradient of light in the dish which makes positively heliotropic animals move to the light, but that difference in intensity on both sides of the animal which is caused by the screening effect of the animal's own body. The same holds true for chemotropism.

Thus far we have discussed positively heliotropic

HELIOTROPISM

animals only. In the case of unequal illumination of the two eyes or of the two sides of the body of a negatively heliotropic animal the tension in the muscles turning the animal to the source of light is diminished. The impulses for locomotion which are equal for the muscles of both sides of the body will, therefore, result in turning the head of the animal away from the source of light. As soon as the plane of symmetry of the animal goes again through the source of light, the symmetrical photosensitive elements of the head receive again equal illumination, and the animal will now continue to move in a straight line away from the source of light. The fully grown larvæ of the housefly when they are ready to pupate show this negative heliotropism.

Negatively heliotropic animals, *e.g.*, the fully grown larvæ of the blowfly, can be made to move from weak light to strong light, *e.g.*, from the shade into direct sunlight, if in so doing the illumination on the two sides of the body remains equal.[287] This was shown by the writer by an arrangement similar in principle to the one described above. Thus the idea that the intensity gradient of light determines the direction of motion was disproved also for negatively heliotropic animals.

Thus far we have shown only that a heliotropic animal is oriented in such a way to a source of light that its plane of symmetry goes through the source of light. This does not yet explain why a positively heliotropic animal cannot go away from the source of light, since in going to or going away from the source of light both sides of the animal receive equal illumination. The fact that a positively heliotropic animal cannot go away from the light finds its explanation by observations of Holmes [228] and Garrey,[177] showing that when light falls from behind and above on a positively heliotropic animal its progressive motions are stopped, and in some cases a tendency to turn a somersault backward may even arise. The case is similar to that of galvanotropism when the current goes through an animal lengthwise (see previous chapter). We must conclude from the observations of Holmes and Garrey, which will be discussed farther on, that if the head of a positively heliotropic animal is turned to a source of light its forward motions are facilitated and the backward motions rendered difficult; while in the case of a negatively heliotropic animal it is just the reverse. If the animal now moves to the right or to the left the illumination of the two eyes or of the two sides of the body becomes different again, causing a forced movement, whereby the plane of symmetry of the moving animal is caused to go through the source of light again; with the head toward the source of light when the animal is positively heliotropic or away from it when it is negatively heliotropic.

2. DIRECT PROOF OF THE MUSCLE TENSION THEORY OF HELIOTROPISM IN MOTILE ANIMALS

The fact that light causes forced movements, like those described in the case of brain lesions and of galvanotropism, has been proved by many observers, and especially clearly by Holmes and Garrey. Holmes worked on the positively heliotropic water scorpion *Ranatra*.[228] When this animal is illuminated from the right side, the legs on the right side of the body are bent and those on the left side extended (Fig. 17). This effect is identical with the one observed in *Palæmonetes,* when a galvanic current goes sidewise through the animal. Hence *Ranatra*

HELIOTROPISM

can easily move to the source of light on its right side but with difficulty or not at all in the opposite direction.

When the light is placed behind the animal, the body is raised up in front and the head held high in the air (Fig. 18). The opposite attitude is assumed, when the light is placed in front, the body being lowered and the head bent down (Fig. 18). These effects resemble the

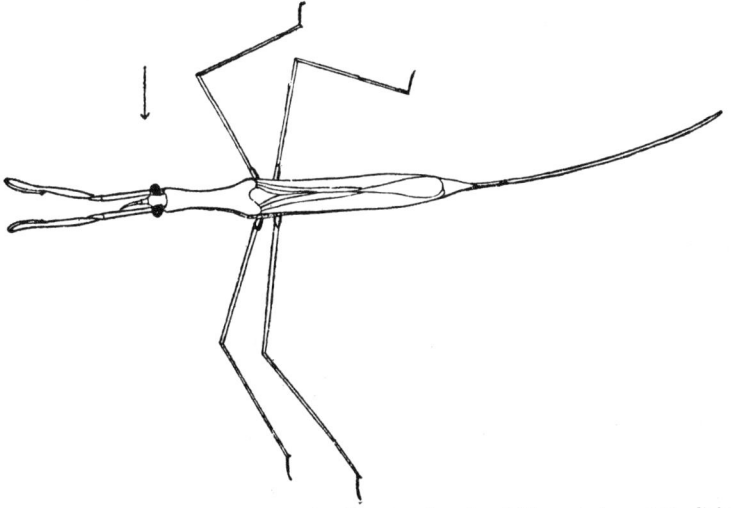

Fig. 17.—Position of the water scorpion *Ranatra* when the right eye is toward the light. (After Holmes.)

galvanotropic effects observed in the position of the head of *Amblystoma* when the current goes forward or backward through the animal.

These latter observations of Holmes explain, as already mentioned, why a positively heliotropic animal cannot move away from the light and why a negatively heliotropic animal cannot move to a source of light. The progressive motions of the negatively heliotropic animal will be stopped when the light strikes it in front; while these motions of the positively heliotropic animal will be facilitated when the light is in front and will be rendered impossible when the light is behind.

The writer had observed long ago that when the convexity of one eye is cut off in the housefly it will no longer go in a straight line but will make circus movements, the normal eye being directed toward the center of the circle.[286] It was shown by Parker that blackening of one

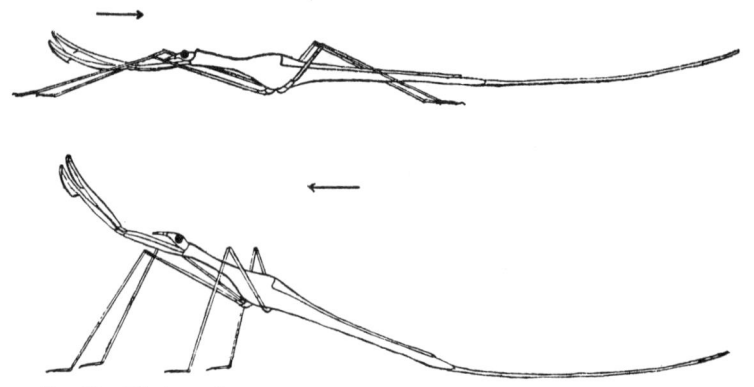

Fig. 18.—The lower figure represents the position of *Ranatra* when the light is behind the body. The upper figure represents the position assumed when the light is in front of the animal. (After Holmes.)

eye of the positively heliotropic butterfly *Vanessa antiopa* calls forth circus movements, with the unblackened eye toward the center of the circle.[398] Holmes, Rádl,[447] Axenfeld, Garrey,[177] and many other authors have since made similar observations. In the positively heliotropic *Ranatra,* Holmes described the effect of blackening one eye as follows:

If one eye of *Ranatra* is blackened over or destroyed the insect in most cases no longer walks in a straight line but performs more or less decided circus movements toward the normal side. Under the stimulus of light the insect assumes a peculiar attitude; the body leans over toward the normal side and the head is tilted over in the same direction.[228]

HELIOTROPISM 55

This is the combination of circus movements with rolling movements familiar to those who have experimented on the brain of fish, where a destruction of one side of the midbrain calls forth rolling motions as well as circus motions toward the same side. Holmes's observations

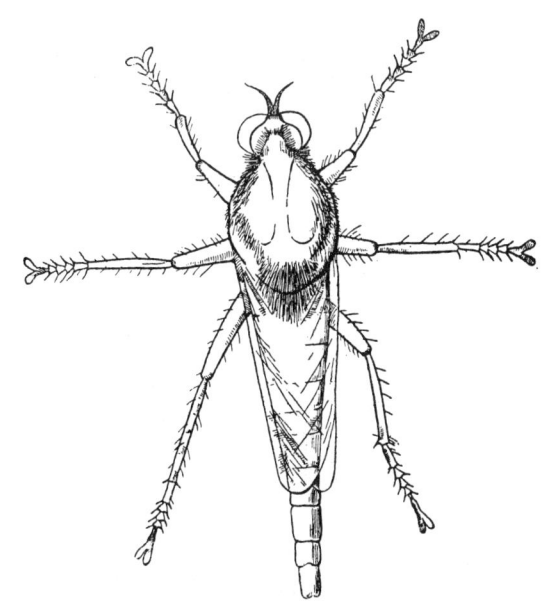

Fig. 19.—Robber fly under normal conditions seen from above. (After Garrey.)

were extended by Garrey's experiments on a large number of insects. Garrey found that the robber fly (*Procta-canthus*) (Fig. 19), which is positively heliotropic, is an unusually good object for the demonstration that the heliotropic reactions of animals are of the type of forced movements. When one eye of this fly is blackened the legs on the side of the unblackened eye are flexed and the legs on the side of the blackened eye are more extended than normally and spread farther apart.[a] The body may tilt as far toward the side of the unblackened eye as to press the legs to the table (Fig. 20). There is sometimes a

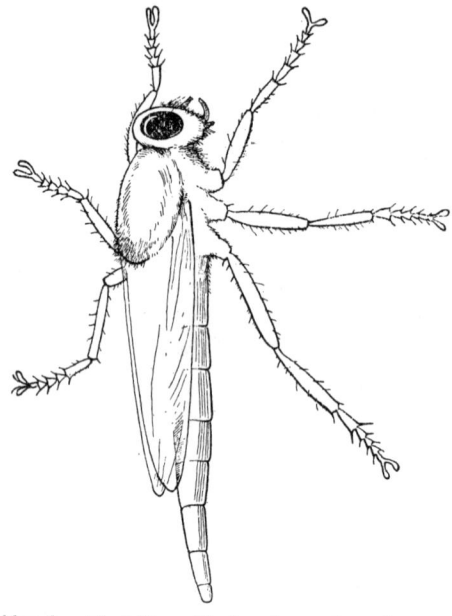

Fig. 20.—Robber fly with right eye blackened, seen from above as in Fig. 19. The body tilts over to the left side so that only the right eye is visible from above. Position of legs changed in such a way as to make motion toward left possible, toward right impossible. (After Garrey.)

tendency on the part of the body of the animal to become slightly concave toward the side of the unblackened eye.

Garrey found also that the same changes take place when one eye receives a stronger illumination than the

---

[a] Figs. 19 to 22 and 27 were drawn from photographs kindly given to the writer for this purpose by Professor Garrey. The draughtsman was unfortunately not familiar with the anatomy of insects, which accounts for shortcomings in the drawings, which, however, have no bearing on the problem for which the drawings are intended.

other. Bringing one eye into the bright beam of light directed through the objective of the optical system of the string galvanometer, while the other eye is illuminated only by the subdued light of the optical room, promptly produced the same changes in the position of the legs and body which were observed when one eye was blackened, the more weakly illuminated eye acting like the blackened eye in the former experiment. When the illu-

FIG. 21.—Position of robber fly when the lower halves of both eyes are blackened. Head tilted up. (After Garrey.)

mination on one side of such animals is stronger than on the other the legs on the more strongly illuminated side of the animal are bent, those on the opposite side are extended; and the head has a tendency to bend toward the light. When an impulse to move originates in the animal, it can turn easily to the light but with difficulty in the opposite direction. As soon as its head is turned to the source of light and both eyes receive the same illumination the difference in tension of the legs on the two sides of the body disappears and now the animal moves or is carried in a straight direction toward the light. By these experiments the proof of the writer's muscle tension theory of heliotropism is made complete.[177]

Garrey observed that when the lower halves of the eyes of the robber fly are blackened the position of the legs of the two sides is symmetrical, but the anterior and middle pairs of legs are extended forward to the maximal extent, producing a striking posture in which the anterior end of the robber fly is pushed up and back from the surface of the table. The body is in opisthotonus with the abdomen concave on the dorsal side, while the head is tilted far up and back (Fig. 21).

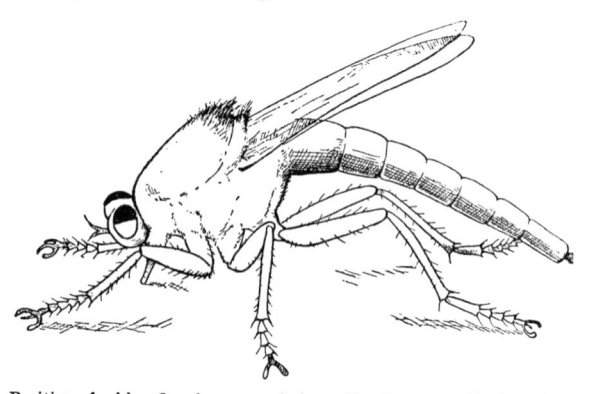

Fig. 22.—Position of robber fly when upper halves of both eyes are blackened. Head down, body convex above. (After Garrey.)

When walking these robber flies gave the impression of trying to climb up into the air. The wings are frequently somewhat spread and the animal may push itself up and back until poised vertically on the tips of the wings and abdomen. The tendency to fly is very pronounced in this condition and upon the slightest disturbance the fly soars upward and backward, striking the top of a confining glass dish or completing a circle by "looping the loop" backward. If it falls upon its back it rights itself by turning a backward somersault. Unequal blackening of the lower parts of the two eyes results in a combination of the effects just described, with those described for blackening one eye, for the animal also performs circus motions.

With the upper halves of the eyes blackened the attitude is the reverse of that described in the preceding section (Fig. 22). The anterior and middle pairs of legs are flexed. The anterior and posterior

HELIOTROPISM

ends of the body bend ventrally with the body in emprosthotonus. The head is bent far down. The animal may actually stand on its head, but the abdomen retains its ventral curvature, leaving a considerable angle open between its dorsum and the wings which normally rest on it.

In both walking and flying it continually keeps close to the table, and upon encountering an obstacle it frequently does a forward somersault. If it gets on its back it rights itself with greatest difficulty as its efforts simply result in bending the tail and head ventrally until they may form a complete ring. In galvanotropism the same general picture is presented by *Palæmonetes* and *Amblystoma* when the anode is at the head end, the tonus changes involved being identical in the two conditions (Garrey [177]).

These experiments leave no doubt that the primary effect of light consists in changes in the tension of muscles and that the heliotropic reactions which appeared to the older observers as voluntary acts are in reality forced movements.

In the chapter on forced movements after brain lesion the fact was mentioned that a dog which had shown circus movements to the left after lesion of the left cerebral hemisphere shows circus motions to the right when afterward the right hemisphere is injured symmetrically; instead of being a physiologically symmetrical animal again after the second operation. The explanation is that the new operation is more effective than the old one whose effect has partly worn off. Garrey has made an observation on heliotropism which shows some analogy with this experiment on the brain.

He found [177] that "robber flies with one eye blackened show the postural conditions in the most pronounced way in the early morning or after being kept for some hours in the dark. Constant exposure to the light produces considerable fatigue of the eye with recovery in the dark. These facts among others suggested the possibility of producing a different sensitiveness of the two eyes and corresponding differences in the muscle tonus with asymmetry of position, and in physiological action of the muscles of the two sides of the body when the two eyes were equally illuminated. Such an experiment constitutes a crucial test of the tonus theory of heliotropism. It succeeded beyond our greatest expectations. Asphalt black was applied to the right eye of several specimens of *Proctacanthus*. In two or three days the paint had formed a brittle shell. During this time the blackened eye had become 'dark adapted.' When such a fly is exposed to light, it tilts and circles to the left. If now the brittle shell is cracked off the right eye by carefully pinching with fine forceps, the exposure of this very sensitive eye to light results in a reversal of the whole picture; the fly circles toward the side from which the black was removed. Although the illumination of the two eyes is of equal intensity, what was the normal eye now becomes relatively a darkened eye owing to its lesser sensitiveness. A differential effect results, probably due to a difference in the rate of photochemical change in the two eyes. This reversal of the muscle tonus and of forced motions may persist for an hour or two or even longer, until the two eyes become, as they ultimately do, of equal sensitiveness and the fly behaves like a normal animal.

"These experiments are not only incompatible with any 'avoidance' idea, for after removal of the black there is nothing to avoid, but they are also incompatible with the conception of 'habit formation,' for 'habit' in the performance of the circling movements is of no avail when light is admitted to the darkened eye—the animals circle to that side because the tonus of the muscles is such that they are forced to do so.

HELIOTROPISM

"All the experiments show that the muscle tone is dependent upon the intensity of the light and that the postures assumed depend upon the relative difference in the light stimulus to the eyes. In animals with one eye completely covered the radii of the circles in which they moved were shorter the more intense the illumination of the normal eye. With one eye partially covered the circles were larger than when completely covered, and in the same way the circles were larger when one eye was covered by a film of collodion or of brown shellac, which admits some light, than when subsequently covered by opaque asphalt black. When one eye was partially covered by central application of the black paint the tilting and circling to the opposite side were abolished or reversed by brilliant illumination of the partially blackened eye. These results explain why a positively heliotropic animal with one eye blackened approaches a light by a series of alternating small and large circles, the former being executed when the good eye is illuminated from the source of light, the larger when it is in the shadow."

We have thus far discussed chiefly positively heliotropic animals, *i.e.*, animals which are compelled to move toward the source of light. The difference between these and negatively heliotropic animals is that the legs on the illuminated side of a negatively heliotropic animal are extended, while those on the opposite side are in flexed position. This has been directly observed by Holmes, who also made sure of the fact that negatively heliotropic animals, when one eye is blackened, turn in circles with the blackened eye toward the center of the circle [228]; while positively heliotropic animals turn in circles with the unblackened eye toward the center of the circle.

3. HELIOTROPISM OF UNICELLULAR ORGANISMS

In unicellular organisms, where cilia act as locomotor organs, it can easily be shown that the orientation by light is of the nature of changes in the position of cilia; this is for instance the case in respect to *Volvox*. Holmes[226] states for the heliotropic reactions of this organism, that they are due to differences in the activity of the cilia on both sides of the organism and this explanation agrees with the actual observations of Bancroft on the galvanotropic reactions of *Volvox*.

In flagellates, the mechanism of locomotion is very complicated and does not consist in an oar-like action of a flagellum as was formerly assumed. Bancroft has shown that in *Euglena*, as already stated, the flagellum inserted at the anterior end of the organism is bent backward in the form of an inverted U, and that locomotion is brought about by the formation of a loop which travels from the base of the flagellum toward the free end (Fig. 23). The path of the organism which results from this action is a spiral with continual rotation of the organism around its longitudinal axis. Bancroft has shown that the behavior of the organism under the influence of light is identical with that in a constant galvanic field.[21] One-sided illumination as well as a current going transversally through such an organism cause changes in the position of cilia comparable with those observed in the legs of crustaceans, insects, and vertebrates.

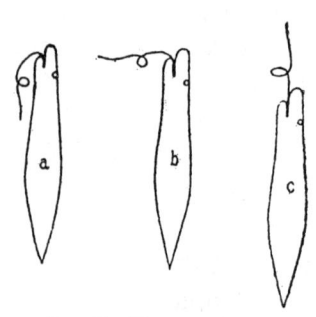

FIG. 23.—Diagram showing the position of the flagellum as seen in a viscid medium. *a*, when *Euglena* is swimming forward in a narrow spiral; *b*, when swerving sharply toward the dorsal side; *c*, when moving backward. (After Bancroft.)

HELIOTROPISM

4. Heliotropism of Sessile Animals

When we study the effects of light on sessile animals we find that they behave in a similar manner to sessile plants. When illuminated from one side they bend their heads to the source of light until their axis of symmetry goes through the source of light. In this case the symmetrical photosensitive elements receive equal illumination and the symmetrical muscles are under equal tension. Hence the animal remains in this orientation. These sessile animals were the first examples by which the

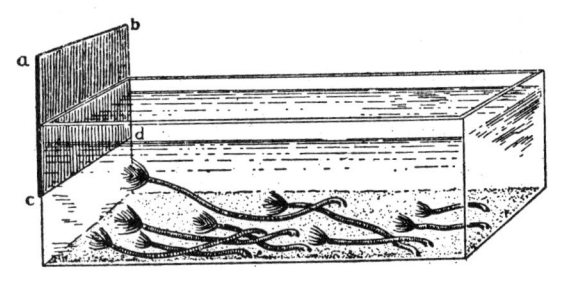

FIG. 24.—Tube worms in aquarium, all bending toward light.

muscle tension theory of animal heliotropism was proved.[288]

Spirographis spallanzani (Fig. 24) is a marine annelid from 10 cm. to 20 cm. long, which lives in a rather rigid yet flexible tube. The latter is formed by a secretion from glands at the surface of the animal. The tube is attached by the animal with its lower end to some solid body, while the other end projects into the water. The worm lives in the tube and only the gills, which are arranged in a spiral at the head end of the worm, project from the tube. The gills, however, are quickly retracted, and the worm withdraws into the tube when touched or if a shadow is cast upon it.

When such tubes with their inhabitants are put into an aquarium which receives light from one side only, it requires, as a rule, a day or more until the foot end of the tube is again attached to the bottom of the aquarium. As soon as this occurs, the anterior end of the tube is raised by the worm until the axis of symmetry of the gills falls into the direction of the rays of light (Fig. 24) which enter through the window into the aquarium.[288] When the animal has once reached this position it retains it as

Fig. 25.—The same animals after the position of the aquarium to the window was reversed.

long as the position of the aquarium and the direction of the rays of light remain the same. When, however, the aquarium is turned 180°, so that the light falls in from the opposite direction, the animal bends its tube during the next twenty-four or forty-eight hours in such a way that the axis of symmetry of its circle of gills is again in the direction of the rays of light (Fig. 25). When the light strikes the aquarium from above, the animals assume an erect position, like the positively heliotropic stems of plants when they grow in the open.

In these phenomena the mechanical properties of the tube play a rôle. When the animal is taken out of the bent tube, the latter retains its form. This permanent change of form of the tube is apparently caused through the secretion of new layers on the inside of the tube. The youngest layers of the secretion are more elastic than the old layers, and, moreover, have at first a powerful tendency to shorten. If such a secretion occurs on one side of the tube only, or more so than on the opposite side, the former must become shorter than the latter, and the result must be a curvature of the tube, that side becoming concave where the new secretion has occurred.

On this assumption the process of heliotropic curvature is in this case as follows: when the light strikes the circle of gills from one side only, in these elements certain photochemical reactions occur to a larger extent, than on the opposite side. This results in corresponding alterations of the sensory nerve endings, the sensory nerves and the corresponding motor nerves, and their muscles. The sense of these changes is such as to throw the muscles connected with the nerves of the gills on the light side into a more powerful tonic or static contraction than the muscles on the opposite side of the body. The consequence is a bending of the circle of tentacles, or the head, toward the source of light, which will continue until the axis of symmetry of the circle of tentacles falls into the direction of the rays of light. When this happens, symmetrical tentacles are struck at the same angle (or in other words, with equal intensity) by the rays of light, and therefore the tone of the antagonistic muscles is the same. The result is that the circle of tentacles becomes fixed in this position. The bending of the head produces an increased pressure and friction of the animal against that side of the tube which is directed toward the light, and this pressure and friction lead to an increased secretion and the formation of a new layer inside the tube.

Heliotropic curvature of sessile animals can be shown equally well in a hydroid, *Eudendrium*. It is necessary to cut off the old polyps at once when the animal is brought into the laboratory and to put the stem into fresh, clear,

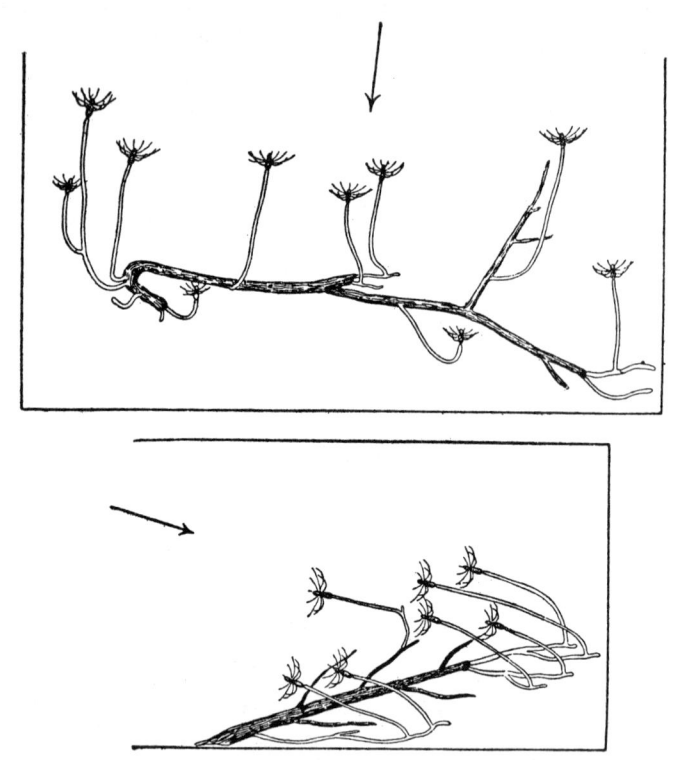

FIG. 26.—Polyps of *Eudendrium*, all growing toward source of light. The arrow indicates the direction of the rays of light, which in one case fall in from above, in the other from the left side.

sea water. In a day or two new polyps are formed by regeneration and these polyps will bend toward the light until their axis of symmetry is in the direction of the rays of light (Fig. 26). The region at the base of the polyps is contractile and when light strikes the polyps from one

HELIOTROPISM

side only, the stem on the side of the light contracts more than on the other side, and this results in a bending of the stem, whereby the polyp is put into the direction of the rays of light. As soon as the axis of the polyp is in the direction of the rays of light (provided there is only one source of light), the tension of the contractile elements is the same all around, and there is no more reason for the organism to change its orientation. It, therefore, remains in this orientation to the light.

The muscle tension theory of animal heliotropism is, therefore, proved for all classes of the animal kingdom, infusorians, hydroids, annelids, crustaceans, etc. It would be wrong to state that the theory holds only for insects.

CHAPTER VI

AN ARTIFICIAL HELIOTROPIC MACHINE

THE reader will have perceived that in the preceding analysis animals are treated as machines whose apparently volitional or instinctive acts, as *e.g.*, the motion toward the light, are purely physical phenomena. The best proof of the correctness of our view would consist in the fact that machines could be built showing the same type of volition or instinct as an animal going to the light. This proof has been furnished by the well-known inventor, Mr. John Hays Hammond, Jr. The following is a description of the machine given by one of Mr. Hammond's fellow-workers who coöperated with him in the development of the machine, Mr. B. F. Miessner.

This "Orientation Mechanism" consists of a rectangular box, about 3 feet long, 1½ feet wide, and 1 foot high. This box contains all the instruments and mechanism, and is mounted on three wheels, two of which are geared to a driving motor, and the third, on the rear end, is so mounted that its bearings can be turned by solenoid electro-magnets in a horizontal plane. Two 5-inch condensing lenses on the forward end appear very much like large eyes.

If a portable electric light, such as a hand flashlight, be turned on in front of the machine it will immediately begin to move toward the light and, moreover, will follow that light all around the room in many complex manoeuvres at a speed of about 3 feet per second. The smallest circle in which it will turn is about 10 feet diameter; this is due to the limiting motion of the steering wheel.

Upon shading or switching off the light the " dog " can be stopped immediately, but it will resume its course behind the moving light so long as the light reaches the condensing lenses in *sufficient intensity*. Indeed, it is more faithful in this respect than the proverbial ass behind the bucket of oats. To the uninitiated the performance of the pseudo dog is very uncanny indeed.

The explanation is very similar to that given by Jacques Loeb, of reasons responsible for the flight of moths into a flame. . . .

The *orientation mechanism* here mentioned possesses two selenium cells corresponding to the two eyes of the moth, which when influenced by light effect the control of sensitive relays instead of controlling nervous apparatus, as is done in the moth. The two relays (500 to 1,000 ohm polarized preferred) controlled by the selenium cells in turn control electro-magnetic switches, which effect the following operations: When one cell or both are illuminated the current is switched on to the driving motor; when one cell alone is illuminated an electro-magnet is energized and effects the turning of the rear steering wheel. The resultant turning of the machine will be such as to bring the shaded cell into the light. As soon and as long as both cells are equally illuminated in sufficient intensity, the machine moves in a straight line toward the light source. By throwing a switch, which reverses the driving motors, the machine can be made to back away from the light in a most surprising manner. When the intensity of the illumination is so decreased by the increasing distance from the light source that the resistance of the cells approach their dark resistances, the sensitive relays break their respective circuits and the machine stops.

The principle of this orientation mechanism has been applied to the "Hammond Dirigible Torpedo" for demonstrating what is known as *attraction by interference*. That is, if the enemy tries to interfere with the guiding station's control the torpedo will be attracted to him, etc.[a]

Nothing seems to have been published beyond these meagre details, but the writer understands that the active machine has been demonstrated in a number of places in this country. It seems to the writer that the actual construction of a heliotropic machine not only supports the mechanistic conception of the volitional and instinctive actions of animals but also the writer's theory of heliotropism, since this theory served as the basis in the construction of the machine. We may feel safe in stating that there is no more reason to ascribe the heliotropic reactions of lower animals to any form of sensation, *e.g.*, of brightness or color or pleasure or curiosity, than it is to ascribe the heliotropic reactions of Mr. Hammond's machine to such sensations.

---

[a] *Electrical Experimenter*, September, 1915, 202.

CHAPTER VII

ASYMMETRICAL ANIMALS

It was necessary for us to begin our analysis with symmetrical animals since as the result of this analysis the conduct of asymmetrical organisms offers no difficulty. The result of the asymmetry consists merely in a change in the geometrical character of the path in which an animal is compelled to move to or from the source of energy. While this path is a straight line in a symmetrical and positively heliotropic organism it is a spiral around this straight line as an axis in an asymmetrical organism, like *Euglena*. Suppose a positively heliotropic animal to have slightly asymmetrical appendages which give it a tendency to deviate to the left. Let us suppose that the plane of symmetry of the animal goes at the beginning of the experiment through the source of light and that the animal is swimming toward the light. After a few strokes the head of the organism will have deviated slightly to the left on account of the asymmetry in the activity of the appendages. As soon as the median plane of the animal deviates to the left, the left eye is less illuminated than the right one. As a consequence, a difference in the tension of the muscles on the two sides of the animal will be produced which will compensate the natural lack of symmetry in the muscles and the animal will cease to deviate further to the left; and this compensating effect of the unequal illumination of the two eyes will continue until the animal is actually oriented in the right way again, *i.e.*, until its plane of symmetry goes through the source of light. All that the inherited or accidental asymmetry does is to cause the animal to move in a path which is not a mathematically straight line; but this deviation will be marked only in a case of very pronounced or excessive asymmetry.

We have already described the behavior of a dog whose left cerebral hemisphere has been injured and who has a tendency to deviate to the left. When such a dog is shown a piece of meat it moves toward it in a fairly straight line, its tendency to deviate to the left being compensated by the orienting effect of the retina image of the piece of meat. If the dog deviates to the left, the piece of meat is apparently dislocated to the right of the dog and this dislocation alters the tension of the muscles on the two sides of the animal in such a way as to make it turn back to the right. In this way the dog reaches the piece of meat in a fairly straight line, though with a greater amount of labor, since the tendency to deviate to the left is constantly compensated automatically by a stronger contraction of the muscles turning the animal to the right.

The writer showed many years ago that many insects have a tendency to creep upward, and that this is due to an orienting effect of gravity upon the animal. When a perfectly symmetrical insect is put on a vertical stick it walks upward in a straight line. What will happen when such an animal is made asymmetrical? Garrey has performed this experiment by using flies in which one eye was blackened. As we have seen, such organisms are rendered asymmetrical not only in regard to the eyes but also in regard to their apparatus of locomotion, since in one side of the body the tension of the flexors, in the legs of the other side the tension of the extensors prevails. As a consequence the fly has a tendency to move in circles with the intact eye toward the center.

Garrey has shown that when a fly with one eye blackened is put on a vertical stick, it still walks upward, but in spirals around the stick (Fig. 27), instead of in a straight line. The asymmetry of locomotion changes only the geometrical nature of the path in which the animal moves, from a straight line to a spiral, but does not alter the forced movement character of the reaction.

Bancroft has pointed out that when in a positively heliotropic amphipod one eye is blackened and the legs of the same side are cut off, the animal's path would be a combination of a circus motion induced by the blackening of the eye and of a rolling motion around its longitudinal axis. Both effects combined would result in the animal swimming in a spiral path, and if the animal is positively heliotropic it would swim in such a path toward the light. This is the path which aquatic, asymmetrical positively heliotropic organisms, such as the flagellate *Euglena,* describe in their motions to the light.

FIG. 27.—Fly with one (right) eye blackened can creep only in a spiral on a vertical stick, while normally it creeps in a straight line. (After Garrey.)

ASYMMETRICAL ANIMALS

But this locomotor mechanism (of *Euglena*) is imperfect, it forces the organism to move in a spiral, and always to turn toward a structurally determined side. There are many organisms which swim in spirals and become oriented by turning toward a structurally defined side. Jennings and Mast include all such orientations under "trial and error" and contrast them with the direct orientation of such animals as the amphipods in which the turning may be either toward the left or the right. Let us now consider whether the orientation of *Euglena* is more like the selection of random movements (which we would all agree may justifiably be called "trial and error"), or whether it is more like the orientation of the terrestrial amphipods studied by Holmes.

I think that all students of behavior including Jennings and Mast believe that in the case of these amphipods we have direct heliotropic orientation. If the right eye of such a positively heliotropic amphipod be covered with asphalt varnish it will execute circus movements towards the left. The usual explanation is that the main nervous connection is between the eye on one side and the legs on the opposite side of the body. The light shining on the uncovered eye brings about a condition of increased muscular tonus in the legs of the opposite side, which is not present in the legs connected with the covered eye. Consequently the right legs push more strongly and the amphipod turns towards the left.

Suppose now we remove some or all of the left legs from an amphipod of this kind so that it will always turn toward the left, and transfer it to water in which it must be supposed to swim in a spiral path. We will then have an organism which would become oriented in essentially the same way that *Euglena* does. The animal would always swerve toward the left. But, when the spiral course brings it into such a position that the light shines directly on the left eye, the muscular tonus of the right legs would be increased and the swerving toward the light would increase. Thus orientation would be effected in just the same way that it is in *Euglena*.

While these hypothetical changes that must be made in the amphipod, to make it react like *Euglena*, are considerable, they concern only the details. The fundamental nature of the photochemical substances, the nature of their stimulation and the character of their connection with the locomotor organs have none of them been modified. All that has been done is to make an asymmetrical organism swimming in a spiral out of a bilateral one.[a] These changes are much less fundamental than

---

[a] Swimming in a straight line.

those which we would have to imagine in order to make an amphipod orient to light by the selection of random movements. In order to bring about this latter change the whole nature of the photochemical substances and their relations to the leg muscles would have to be modified. In the one case the required changes are all of a mechanical nature and so simple that the experiment might possibly succeed. In the other case the required changes are largely chemical, and so complex that we have no data for even imagining what ought to be done in order to bring them about (Bancroft [21]).

The asymmetry of organisms only modifies the geometrical character of the path but not the mechanism of the reaction.

CHAPTER VIII

TWO SOURCES OF LIGHT OF DIFFERENT INTENSITY

THE writer observed that if heliotropic animals are exposed to two equidistant lights of equal intensity they move in a line perpendicular to the line connecting the two lights.[287, 294] This has been confirmed by numerous observers, *e.g.*, Bohn on *Littorina,* by Parker and his pupils, especially by Bradley M. Patten on the larvæ of the blowfly, by Loeb and Northrop on the motions of the larvæ of *Balanus,* and by others. The question arises: In which line will an animal move when the intensity of the two lights differs? When the animal is positively heliotropic it should cease to move in a line at right angles to the line connecting the two lights but should move in a line which deviates toward the stronger of the two lights; if the animal is negatively heliotropic it should deviate toward the weaker of the two lights. When the two eyes are illuminated by two lights of different intensity, the illumination in both eyes can become approximately equal only when the eye struck by the weaker light is exposed at a larger angle than the eye struck by the stronger light. Under such conditions, the animal should be compelled to move in a straight line which, however, is no longer at right angles to the line connecting the two lights, but which deviates to an extent determined by the difference in the intensity of the two lights. The case was

76   TROPISMS

worked out quantitatively by Bradley M. Patten on a negatively heliotropic animal, the full grown larva of the blowfly.[412,413] The source of light was at G (Fig. 28) (one or more Nernst lamps of measured candle power), a portion of light from these lamps passed through the screens d and d' to the mirrors M and M', set at a definite

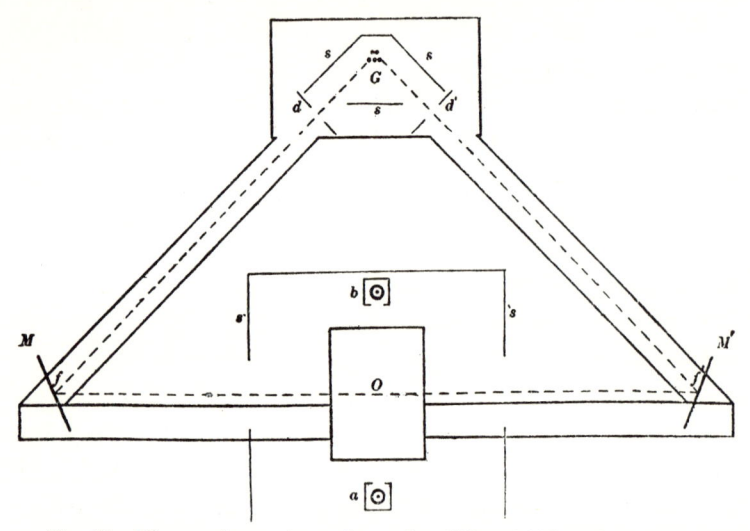

FIG. 28.—Diagram of apparatus used to produce differential bilateral light stimulation. G, five 220-volt Nernst glowers; M and M', mirrors; f and f', central point of mirrors; O, center of observation stage; dotted lines, central ray of beam of light from the glowers reflected to O by the mirrors; d and d', screens with rectangular openings; s and s', light shields; a and b, 2 c.p. orienting light with screens. (After Patten.)

angle so that the rays were reflected to the observation point O. The two beams of light reaching O were of the same intensity. With the means of one of the lights at a or b the animal was first caused to move across the field O at right angles to the rays reflected from the mirrors M and M'. The animals first started in this direction, then came suddenly under the influence of the light reflected by M and M'. In order to make the ratio of intensities of light from M and M' different, the observation stage O was put at unequal distance from M and M'. The larvæ were made to record their trails while moving under the influence of two lights and the deviation of this trail from the perpendicular upon the line connecting the two sources of light M and M' was measured with a goniometer (Fig. 29). The result of the measurements of 2,500 trails, showing the progressive increase in angular deviation of the larvæ (from the perpendicular upon the line connecting the two lights) with increasing differences between the lights, are given in Table I. Since the deviation or angular deflection was toward the weaker of the two lights (the animal being negatively heliotropic) the deviation is marked negative.

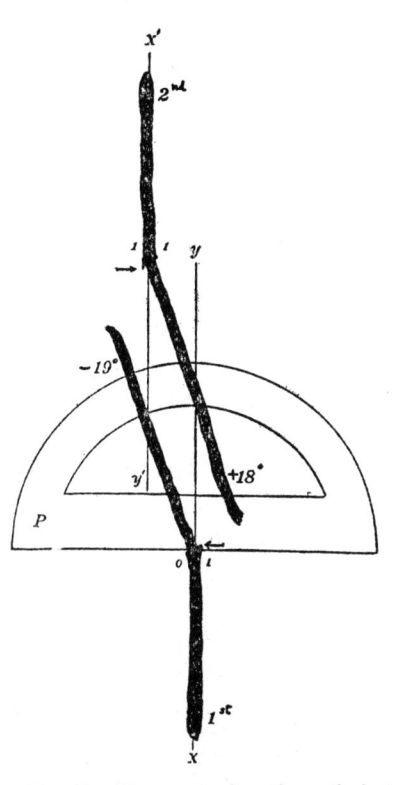

Fig. 29.—Diagram to show the method of measuring trails. The lines xy and $x'y'$ are drawn through the trails at the points reached—marked by the arrows—when the side lights were turned on. The angle of deflection from this line is measured by a protractor, P. The small figures near the arrows indicate the number of wig-wag movements made when the side lights were turned on; 1st and 2nd refer to the sequence in which the trails were run. (After Patten.)

TABLE I.

Percentage difference in the intensity of the two lights	Average angular deflection of the two paths of the larvæ toward the weaker light
Per cent.	Degrees
0	− 0.09
8⅓	− 2.77
16⅔	− 5.75
25	− 8.86
33⅓	−11.92
50	−20.28
66⅔	−30.90
83⅓	−46.81
100	−77.56

Patten also investigated the question whether the same difference of percentage between two lights would give the same deviation, regardless of the absolute intensities of the lights used (Weber's law). The absolute intensity was varied by using in turn from one to five glowers. The relative intensity between the two lights varied in succession by 0, 8 1/3, 16 2/3, 25, 33 1/3, and 50 per cent. Yet the angular deflections were within the limits of error identical for each relative difference of intensity of the two lights, no matter whether 1, 2, 3, 4, or 5 glowers were used. Table II gives the results.

TABLE II

A Table Based on the Measurements of 2,700 Trails Showing the Angular Deflections at Five Different Absolute Intensities

Number of glowers	Difference of intensity between the two lights					
	0 per cent.	8⅓ per cent.	16⅔ per cent.	25 per cent.	33⅓ per cent.	50 per cent.
	Deflection in degrees					
1	−0.55	−2.32	−5.27	−9.04	−11.86	−19.46
2	−0.10	−3.05	−6.12	−8.55	−11.92	−22.28
3	+0.45	−2.60	−5.65	−8.73	−13.15	−20.52
4	−0.025	−2.98	−6.60	−9.66	−11.76	−19.88
5	−0.225	−2.92	−5.125	−8.30	−10.92	−19.28
Average...	−0.09	−2.77	−5.75	−8.86	−11.92	−20.28

TWO SOURCES OF LIGHT

On the writer's theory the following explanation of these deviations should be given. The muscles moving the head of the animal to the side of the weaker illumination, having a higher tension than their antagonists, bring about a deflection of the animal toward the side of the weaker light. As soon as its two photosensitive areas in the head—the animal has no eyes—which are not parallel, but inclined to each other are deflected from the perpendicular upon the line connecting the two lights, the photosensitive areas of the animal will no longer be struck by the lights at the same angle, but on the side of the weaker light the area will be struck at an angle nearer to 90° than the photosensitive area exposed to the stronger light. In this way the change in angle will compensate the difference in intensity of the two lights until the orientation of the animal is such that the compensation is complete and both photosensitive areas receive the same illumination. The animal will then continue to move in this direction.

Patten has computed the angle of the photosensitive surfaces for these animals from the angle of their orientation under varying inequalities of illumination.

This angle has been computed for the blowfly larva, using the "angular deflections" already ascertained. The magnitude of the angle may bear no direct relation to the actual angle at which the sensitive areas are located in the body of the animal, because of the many factors which may modify the direction of the rays before they fall on the sensitive surfaces. The significant test of the hypothesis would be the constancy of the angle when computed from experimental data obtained under varying conditions.

The method of constructing such an angle is shown in Fig. 30, in which the opposing lights are assumed to be of a two-to-one ratio of intensity. The line AB is drawn perpendicular to the direction of the rays of light. On the line AB, construct angle BOC equal to the actual average angular deflection of the larvæ at a two-to-one ratio of lights.

80   TROPISMS

The problem now resolves itself into the construction of an angle about *OC* as a bisector, which shall be of such a magnitude that equal dis-

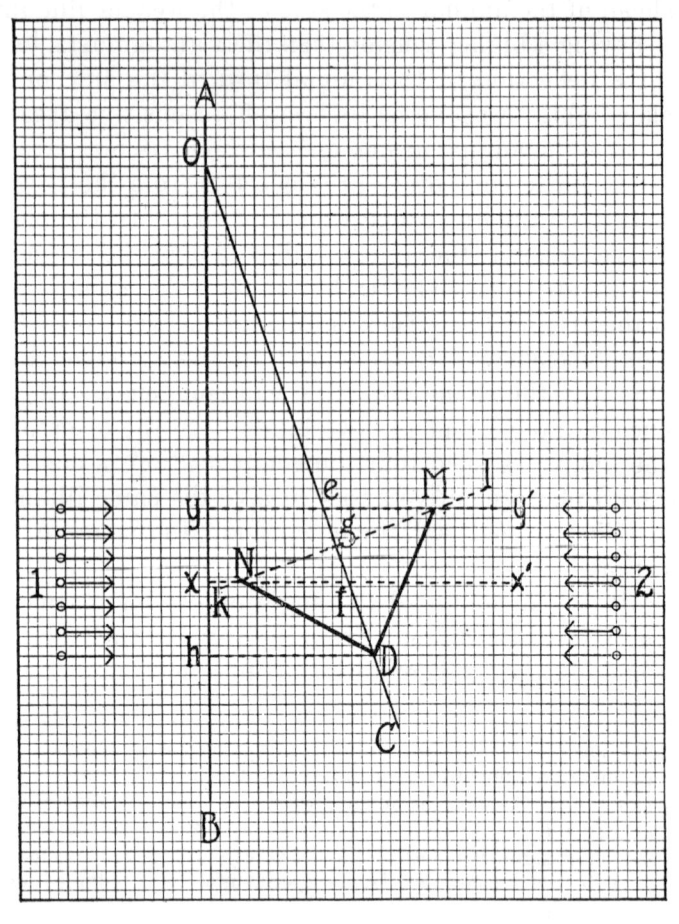

FIG. 30.—Diagram for constructing direction of motion of larvæ under influence of two lights of different intensity. (After Patten.)

tances on its opposite sides shall have projections on the line *AB* of the ratio of two to one.

Construction: From a point *D* on the line *OC* draw *Dh* perpendicular

TWO SOURCES OF LIGHT

to AB. Lay off on AB distances hx and hy, such that $hy = 2hx$. From x and y erect lines perpendicular to AB; they will intersect OC at f and e respectively. Bisect the line ef, and at its middle point, g, construct a line kl perpendicular to OC. From the point of intersection of kl and yy' (M), draw a line to D. From the intersection of kl and xx' (N), draw a line to D.

The angle MDN is the desired angle.

Proof: $eg = gf$ (construction).

Angle egM = angle fgN (construction).

Angle Meg = angle Nfg (alternate int. angles of parallel lines, yy' and xx' being parallel by construction).

Therefore triangle Meg = triangle Ngf (side and two adjacent angles being equal).

$Ng = gM$ (similar sides of equal triangles).

$gD = gD$ (identical).

Therefore triangle NgD = triangle MgD (rt. triangles, altitude and base equal).

Therefore angle gDM = angle gDN and side DM = side DN.

Now by construction hx is the projection of DN on AB and hy the projection of MD on AB, and by construction $hy = 2hx$.

This fulfills all the conditions of construction.

The equal lines MD and DN represent equal bilateral sensitive areas inclined to each other at such an angle, MDN, that the surface represented by MD intercepts an area of light twice as great as the surface represented by DN, its projection on the perpendicular to the light rays being twice as great ($hy = 2hx$). But the light falling on DN is of twice the intensity of the light falling on DM, so that the total amount of light received by each of the equal areas is the same.

By this method of construction, the average angle of sensitiveness was computed for four intensity differences, using as a basis the angular deflection of the larvæ as determined by experiment. The magnitude of the angles is almost identical in all four cases.[412]

Experiments by a somewhat different method, to be discussed in the next chapter, on the positively heliotropic larvæ of the barnacle show that these results of Patten are more general.

We may, therefore, say that the migration of animals to or from the light is of the nature of a forced movement determined by the effect of light on the photosensitive elements of the body. Unequal illumination of symmetrical photosensitive elements on the two sides of the body alters the tension of symmetrical muscles, and as a consequence the animal is, when moving, compelled to change its direction of motion until it is oriented in such a way to the light that symmetrical elements receive the same illumination. In this case the tension of symmetrical muscles is equal again and the animal is compelled to move in this direction.

It has been suggested by the anthropomorphic interpreters of animal conduct that the motion of an animal to a source of light is the same phenomenon as when a human being who has lost his way in the dark is attracted by an illuminated human habitation. As Bohn pointed out, the definite path in which a positively heliotropic animal moves when under the influence of two lights, shows that the anthropomorphic interpretation is as erroneous in this as in any other case. A human being would go to one of two illuminated houses and not toward a point between them, determined by the relative intensity of the two lights.[66]

CHAPTER IX

THE VALIDITY OF THE BUNSEN-ROSCOE LAW FOR THE HELIOTROPIC REACTIONS OF ANIMALS AND PLANTS

We have thus far said little about the identity of the heliotropism of plants and animals. Yet the two phenomena are essentially alike. When we keep positively heliotropic sessile plants and sessile animals near a window, both will bend toward the source of light, though the mechanism of bending may not be the same in all details, the bending being produced in the case of the plant (and possibly in certain animals like *Eudendrium*) by unequal growth in length of the plant on the illuminated and shaded sides; while in the case of higher animals, *e.g.*, *Spirographis*, it is produced by differences in the tension of the muscles on the illuminated and shaded sides of the animal. Motile plant organisms like *Volvox*, are driven to the source of light, owing to differences in the tension of the contractile organs on the shaded and illuminated side, and the same is true for animals like insects.

A further point of coincidence lies in the validity of the photochemical law of Bunsen and Roscoe for the heliotropism of animals and plants.

The law of Bunsen and Roscoe says that within certain limits the chemical effect produced by light increases in proportion with the product of intensity into the duration of illumination, *e.g.*, $Effect = Kit$, where i is intensity, t duration of illumination, and K a constant. This is true for the blackening of photographic paper by light, and it can be shown that the same law holds for heliotropic reactions of plants as well as animals.

Blaauw[46,47] established this fact for the etiolated seedlings of *Avena sativa*. These organisms were exposed to lights of a definite candle power for some time and then left in the dark. After a certain time the seedlings began to bend, becoming concave on that side which had previously been illuminated. By varying the candle power of light (i) and the duration of illumination (t), he found that the value of it required to cause 50 per cent. of the seedlings to bend was always the same. Table III gives

TABLE III

Time required for different intensities of light to produce heliotropic curvatures in 50 per cent. of the seedlings of *Avena*

Candle-meter	Duration of illumination	Candle-meter-seconds
0.00017	43 hours	26.3
0.000439	13 hours	20.6
0.000609	10 hours	21.9
0.000855	6 hours	18.6
0.001769	3 hours	19.1
0.002706	100 minutes	16.2
0.004773	60 minutes	17.2
0.01018	30 minutes	18.3
0.01640	20 minutes	19.7
0.0249	15 minutes	22.4
0.0498	8 minutes	23.9
0.0898	4 minutes	21.6
0.6156	40 seconds	24.8
1.0998	25 seconds	27.5
3.02813	8 seconds	24.2
5.456	4 seconds	21.8
8.453	2 seconds	16.9
18.94	1 second	18.9
45.05	2/5 seconds	18.0
308.7	2/25 seconds	24.7
511.4	1/25 seconds	20.5
1,255	1/55 seconds	22.8
1,902	1/100 seconds	19.0
7,905	1/400 seconds	19.8
13,094	1/800 seconds	16.4
26,520	1/1000 seconds	26.5

the time required for different intensities of light varying from 0.00017 to 26,520 candle power to cause 50 per cent. of the seedlings to show heliotropic curvatures. As can be seen, the product it is always approximately 20.

Ewald and the writer [300, 305] tested the validity of the law of Bunsen and Roscoe for the heliotropic curvatures of *Eudendrium*. A number of stems of *Eudendrium*, from which the polyps had been cut off, were put upright into a trough with parallel walls, containing sea water. As soon as the new polyps had regenerated they were exposed to light of a certain intensity for a short time and then kept in the dark. In the dark the bending of the polyps in the direction of the former source of light occurred. The purpose was to find the minimum time of exposure required for a given light (40 candle power) to induce 50 per cent. of the polyps to bend to the light (Table IV).

TABLE IV

Duration of illumination	Percentage of polyps bending toward the former source of light					
	Distance of the polyps from the light in meters				
	0.25	0.50	1.00	1.50	2.00
10	65[1]				
15	68				
20	74				
30		42			
35					
40		56			
45		60			
50					
60		60			
90					
120		65	30		
150			48, 50		
180					
240					
300			85	40	
360				40	(15)
420				57	

[1] Very young, abnormally sensitive polyps.

If we calculate from this the value of the product it for different intensities of light we find that it obeys the Bunsen-Roscoe law (Table V).

TABLE V

Distance of polyps from light	Time required to call forth heliotropic curvature in 50 per cent. of the polyps	
	Observed	Calculated according to the Bunsen-Roscoe law
Meters	*Minutes*	*Minutes*
0.25	10	
0.50	between 35 and 40	40
1.00	180	160
1.50	between 360 and 420	360

The material varies considerably so that it is not always possible to induce 50 per cent. of the polyps to undergo heliotropic curvature. For this reason Loeb and Wasteneys [312] repeated these experiments by a somewhat different method.

We confined our experiments to three intensities of light by putting the specimens at distances of 25, 37.5, and 50 cm. from a Mazda incandescent lamp, of about 33 Hefner candles. The times of exposure were adjusted so that on the assumption of the applicability of the Bunsen-Roscoe law the same effect, *i.e.*, the same percentage of polyps bending towards the light should be produced. Thus in some experiments the exposure for the three distances given was 10, 22.5, and 40 minutes respectively, in others, 7, 15.75, and 28 minutes, and so on. The ratios of the percentage of polyps bending toward the light for the three distances should be as 1:1:1. Since the material differed widely in different experiments and in different dishes, it was necessary to compute the averages of a large number of experiments.

The colonies, immersed in sea water, were arranged

in a row in rectangular glass dishes, the stems being inserted in holes made in a layer of paraffin mixed with lamp black as in the previous experiments. The rear side of the dish was also coated with the paraffin lamp black mixture in order to prevent reflection of light from the slightly uneven back surface of the dish.

Table VI gives a summary of the results. The first three columns give the times of exposure for the three

TABLE VI

Times of exposure in minutes			Ratio of per cent. of hydranths bending towards light		
25 cm.	37.5 cm.	50 cm.	25 cm.:37.5 cm.	25 cm.: 50 cm.	37.5 cm.:50 cm.
15		60		1.50	
20		80		1.30	
10	22.5	40	1.20	(3.08)	(2.56)
10	22.5	40	0.94	1.47	1.55
10	22.5	40	1.57	(2.30)	(2.43)
10	22.5	40	1.43	1.04	0.94
10	22.5	40	0.76	1.09	1.47
10	22.5	40	1.05	1.13	0.90
					0.96
10	22.5	40	1.15		0.99
7	15.75	28	0.84	0.62	0.74
7	15.75	28	1.70	0.49	0.58
7	15.75	28	0.85	1.25	1.35
7	15.75	28	(2.09)[1]	0.99	1.08
7	15.75	28	1.14	1.15	0.55
7	15.75	28	0.44	0.92	0.44
7	15.75	28	1.52	0.80	0.61
7	15.75	28	0.59	0.36	0.70
7	15.75	28	0.48	1.07	0.31
7	15.75	28	1.00	0.48	1.80
7	15.75	28	0.69	1.09	0.81
7	15.75	28	1.26	0.85	1.09
7	15.75	28	0.86	1.38	0.85
7	15.75	28	0.70	1.07	1.59
7	15.75	28		0.77	1.24
7	15.75	28		0.60	
Mean.................			1.02	0.99	1.02
Probable error............			±0.01	±0.01	±0.01

[1] Bracketed values being extreme variates are excluded from calculations of the means and probable errors.

distances of the source of light, selected, as stated, on the assumption that the Bunsen-Roscoe law holds. On that assumption the ratio of percentage bent in any two or all three dishes on any one day should equal 1.0. These ratios for each pair of distances of the source of light are given in the three other columns of the table. The percentage bending was only compared in dishes containing material regenerated and exposed on any one day, since only in this case was there any likelihood that the material was in any way uniform, and since only in this case the experiments were carried on at the same temperature and the same conditions of regeneration.

The result was that the observed ratios were as 1.02:0.99:1.02 (with a probable error of ±0.01) while the values calculated on the assumption of the validity of the Bunsen-Roscoe law were as 1:1:1; *i.e.,* the results showed as great an approximation between observed and calculated values as one could expect.

There is a second method for testing the validity of the Bunsen-Roscoe law, based on the use of two sources of light of equal intensity.

If it is true that the heliotropic efficiency of light is determined by the product of intensity, i, into duration of illumination, t, we can alter this product by varying t as well as by varying i.

Rádl had shown that the position of the eye of the fresh water crustacean, *Daphnia,* is determined by the position of a source of light,[447] and Ewald [145] found that by exposing the eye to two different sources of light simultaneously the eye is put into a position determined by the relative intensity of the two lights. When one light remained constant and the intensity of the other light was lowered the position of the eye changed. He now could show that when the duration of illumination of one eye was altered by a rotating opaque disk with one sector cut out, the heliotropic effect on the eye of *Daphnia* was the same as when the intensity i of the same light was reduced to an amount corresponding to the Bunsen-Roscoe law.

Under the influence of two constant lights of equal intensity heliotropic animals move in a direction at right angles to the line connecting the two lights. If the law of Bunsen and Roscoe holds, the effect of a constant light should be diminished if a rapidly rotating opaque disk with one sector cut out be put in front of the light, and the diminution should be equal to the fraction of the arc of the sector. Thus a sector of 90°, which reduces the total duration of illumination to one-fourth, should also reduce the heliotropic effect of the light to one-fourth, and the animal should deviate from the old direction in the direction toward the light without a disk before it. If, however, we lower the intensity of the latter light to one-fourth by doubling its distance we also reduce its heliotropic effect to one-fourth, and now the animal should move again in a line at right angles to the line connecting the two lights.

The following experiments carried out by Loeb and Northrop [309] on the larvæ of the barnacle are perhaps the best proof for the validity of the Bunsen-Roscoe law for animal heliotropism.

These animals are small and can be obtained in large numbers. They were made to collect in the corner of a dish with a little sea water and were then sucked up into a pipette *ef*, Fig. 31, which was blackened with the exception of the opening. When such a pipette is put into a glass dish with parallel walls whose bottom is black (by putting paraffin blackened with lampblack at the bottom of the dish) the larvæ will flow out in a fine stream and swim when they are positively heliotropic in a straight line toward the source of light. They thus form a rather narrow white trail on the dark bottom and it is possible to measure the angle of this trail with the line connecting the two lights. In this way in each observation the average trail of thousands of individuals is measured. By using one constant and one intermittent source of light and comparing the results with those obtained by two constant lights we can test the validity of the Bunsen-Roscoe law.

The method of the experiments was as follows: *abcd* (Fig. 31) is a square dish of optical glass with blackened bottom and containing a

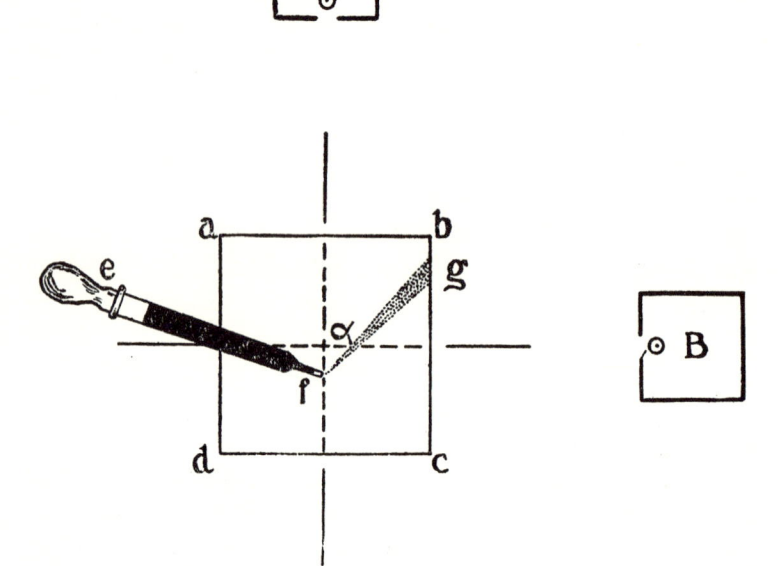

Fig. 31.—Method for the proof of the validity of Bunsen-Roscoe law for the positively heliotropic larvæ of the barnacle. (After Loeb and Northrop.)

layer of sea water. *A* and *B* are two lights, the intensity of which is determined by a Lummer-Brodhun contrast photometer. In front of each light is a screen with a round hole permitting a beam of light to go to the dish. The lights and the dish *abcd* are so adjusted that the two beams of light striking the sides *ab* and *bc* at right angles cross each other in the middle of the dish. The light *A* is fixed while the light *B* is movable on an optical bench. The experiment is made in a dark room and the lights A and B are enclosed in a box. At the beginning of the experiments the pipette is filled with a dense suspension of larvæ in sea water and then put with its point touching the bottom of the dish. The animals flow out in a fine stream which is narrow at the opening of the pipette and widens slightly, owing probably to the negative stereotropism of the animals. A glass plate (Fig. 32) $hikl$, which has a strong red line no and a fine parallel line pq (cut with a diamond), is then put on the dish and so adjusted that pq is in the middle of the stream fg of the animals. Then the angle α which pq makes with the perpendicular from A on ab is measured. This perpendicular is marked in the form of a red line on the black base on which the glass vessel $abcd$ stands. The angle α is measured with a goniometer. When the lights are equal in intensity α should be 45°; if the two lights have different intensities and if A be the stronger light α should become smaller with increasing difference in intensity. The individual measurements vary comparatively little, as long as the difference in the intensity of the two lights is not too great; for this reason our observations do not go beyond a wider ratio of the two lights than 10:1, though 4:1 is perhaps the limit for good results. Table VII gives the results. A is always the stronger light. Each table is the average of from 40 to 60 individual observations, each being the average of the path of many thousands of animals.

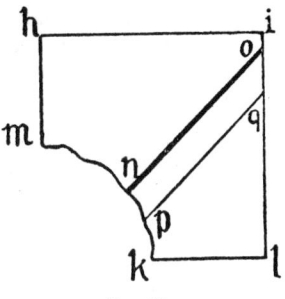

FIG. 32.

TABLE VII

	Value of α for different ratios of intensities of the two lights			
Ratio of the two lights.............	1:1	2:1	4:1	10:1
Value of α (direction of path).......	45.6°	40°	34.4°	28.8°

In the next series of experiments an opaque rotating disk with one sector cut out was placed before light B. In one set of experiments the sector cut out was 90°. The rate of rotation (by an electric motor) was 1,500 to 2,500 revolutions per minute. The other light was constant and its distance was chosen on the assumption of the validity of the Bunsen-Roscoe law for these cases. Thus when the two lights without sector were equal at a given distance of A, by putting 90° sector before

B, it was assumed that the ratio of effects would be the same as if, with constant light, B had been placed at the double distance and the ratio of intensities of the two lights had been 4:1. Going on such a calculation we should expect the same values for α as in Table VII.

As one sees from Table VIII, the observed values are slightly smaller but practically identical with the values obtained when the two lights are constant. The deviation is probably due to the well established fact that the photochemical efficiency of an intermittent light is a trifle less than that calculated on the basis of the Bunsen-Roscoe law.

TABLE VIII

	\multicolumn{3}{c}{Value of α when one light is intermittent (90° sector) and the other constant, and the efficiency of the two lights is calculated on the basis of the validity of the Bunsen-Roscoe photochemical law}
Ratio of the two lights	1:1	2:1	4:1
Value of α	44.2	38.3°	34.1°

We carried out some experiments with a sector of 144°. When the efficiency of both lights was equal on the assumption of the validity of the Bunsen-Roscoe law α was found to be 44.9° (instead of 45°), and for the ratio 2:1 α was found to be 38.8°. The values are, within the limits of error, identical with the values in Tables VII and VIII.[309]

Bradley M. Patten also showed that for the heliotropic reactions of the negatively heliotropic larvæ of the fly the law of Bunsen and Roscoe holds.

Photochemical processes have a very small temperature coefficient and it agrees with this that lowering of temperature within the limits compatible with the motility of animals does not affect the heliotropic response; on the contrary, we shall see that in certain crustaceans (*e.g., Daphnia*) lowering of the temperature may enhance positive heliotropism.[296]

We must, therefore, conclude that the light produces in an eye or an element of the photosensitive skin a chemical reaction which results in the formation of a certain mass of a reaction product. This mass acts on the peripheral nerve endings and brings about an as yet unknown change in the brain elements with which these nerve endings are connected. This change in turn affects the tone or tension of the muscles with which the brain elements are connected. When the rate of photochemical reaction is the same in both eyes or in the photosensitive elements on both sides of the body, the change of tone in the symmetrical muscles of both sides of the body is the same and no change in the position or direction of motion of the organism should occur. If the rate of illumination is different in both eyes, differences in the relative tension of the symmetrical muscles occur, which make the motion to the source of light easy and in the opposite direction more difficult when the animal is positively heliotropic. For the negatively heliotropic animal the opposite effect will be brought about.

These experiments, therefore, show that the tropism theory not only allows us to predict the nature of the animal reactions but allows us to predict them quantitatively. Thus far the tropism theory is the only one which satisfies this demand of exact science.

The degree of directness with which a heliotropic animal goes to or from a source of light depends, aside from the degree of perfection of its locomotor apparatus, upon the intensity of the light and the relative sensitiveness of the animal. Animals which in strong light will move in approximately straight lines to or from the source of light may in weak light reach their goal in a more or less irregular zigzag line. This is easily understood. When an animal by chance gets its median plane too far out of the direction of the rays of light (we assume them to be parallel), the rate of photochemical reaction will become different in both eyes. As soon as the difference between the photochemical reaction products in both eyes exceeds a certain limit the animal will automatically put its plane of symmetry again into the direction of the rays of light. The weaker the light and the less sensitive the animal, the longer it will take until this happens, and the greater the freedom of the animal to deviate from the straight line.

CHAPTER X

THE EFFECT OF RAPID CHANGES IN INTENSITY OF LIGHT

It may prove necessary to make a similar assumption for the effect of a constant illumination as was made by Nernst for the theory of the action of galvanic currents, namely that there are two antagonistic processes going on, one being the photochemical effect of light and the second either a process of diffusion of the substances formed or a chemical reaction of the opposite character as that caused by the action of the light. Many animals which are oriented by constant illumination react by a quick, jerky movement when the intensity of light is either rapidly increased or diminished. In this case the effect is determined by the rapidity of the change in the intensity, $\frac{di}{dt}$, and not by the product of intensity into duration of illumination, it.[297] These twitching or jerking effects caused by a rapidly changing intensity of light are comparable to the twitching brought about in a muscle by a rapid increase or decrease in the intensity of a current. The writer described such reactions first for tube worms like *Serpula,* which withdraws suddenly into its tube when a shadow passes over it or when the intensity of light is suddenly diminished in some other way. The anthropomorphists, of course, declare this reaction to be induced by the instinctive fear of an enemy, oblivious of the fact that if they were consistent they would have to give the same explanation for the twitching of a muscle upon rapid changes in the intensity of a current. The problem to be solved is in both cases a purely physico-chemical one. It was also found that the motions of certain animals stop when they come suddenly from strong light into weak light. This was observed in planarians which as a consequence collect in greater density in spots of the space where the intensity of light is a relative minimum.[291] The difference in the conduct of heliotropic organisms like *Daphnia* which go to or from the light and animals like planarians which come to rest where the intensity of light is a relative minimum can be demonstrated by putting them into a circular vessel (Fig. 33). The positively heliotropic animals collect at *a*, the negatively heliotropic at *b*, while the planarians collect at *c* and *d* where the intensity of light is a minimum. Reactions determined by the value $\frac{di}{dt}$ do not lead to phenomena of orientation, though such (improperly called) "fright reactions"[a] occur in many heliotropic animals; they may lead, however, to collections of animals.

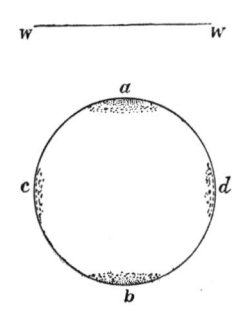

FIG. 33.—Difference in place of gathering between heliotropic animals and animals which come to rest when reaching a relative minimum in the intensity of light. In a circular vessel *a c b d* and *W W* representing the window, positively heliotropic animals will collect at *a*, negatively heliotropic animals at *b*, and animals which come to rest where the intensity of light is a relative minimum at *c* and *d*.

Jennings has maintained that all reactions of unicellular organisms are due to "fright" or "avoiding reac-

---

[a] The reader should notice the difference in the treatment of animal conduct from the point of view of the physicist and of the introspective psychologist. What the physicist expresses correctly by the term $\frac{di}{dt}$ the anthropomorphic biologist explains in terms of human analogy as "avoiding reaction" or "fright reaction," a term which not only assumes the existence of sensations without any adequate proof, but removes the problem from the field of quantitative experimentation.

tions,'' and it seems as if at one time he even intended to deny the existence of tropisms and to maintain that all animals were influenced only by rapidly changing intensities of light. It is needless to discuss such an idea (which he probably no longer holds) in view of the contents of the preceding chapters. He seems, however, to cling to it as far as asymmetrical unicellular organisms are concerned. When moving about *Paramæcia* often reverse the direction of their progressive motion for a moment, but then do not return in the old direction, moving sidewise, on account of the asymmetry in the arrangement of their cilia. Jennings is probably right in assuming that this factor can lead to collections of such infusorians, since it may prevent their leaving a drop and going into the surrounding medium. When, *e.g.*, at the boundary of the two media such a reversal of the action of the cilia occurs, the organisms are prevented from crossing from one medium into the other.

But Jennings goes too far in this attempt, when he tries to explain the heliotropic reactions of certain unicellular organisms, *e.g.*, *Euglena*, in this way. He maintains [253] that unicellular organisms like *Euglena* go to the light on account of shock movements produced by the shading of the photosensitive region of the animal. *Euglena* moves with a constant rotation around its longitudinal axis and Jennings assumes that in a certain phase of the rotation a photosensitive element (the eye spot) of the organism is shaded. This he thinks causes a shock movement, whereby the animal is swerved to the light again during the next half of the spiral revolution, and so on. Similarly in negatively heliotropic *Euglena* the swerving away from the light is, according to Jennings, the shock movement caused by the increased illumination of the photosensitive end of the animal produced by swerving toward the light during the previous half of the spiral revolution. Bancroft [21] showed that Jennings's theory was based upon incomplete facts.

> According to Jennings's view positive heliotropism is conditioned by and should be accompanied by shock movements produced by sudden shading (= shading reaction), and negative heliotropism should always be accompanied by shock movements produced by sudden illumination.
>
> It has been found, however, that this usual association of shock movements with tropism is not a necessary one, but that it can be destroyed if the proper means be taken. Consequently the view that the heliotropic swerving is a shock movement must fall.
>
> When *Euglenæ* from Culture B were placed in the rays of the arc light, at a distance of four or five feet from the light, they were strongly positively heliotropic and gave the shading reaction. When, however, they were gradually brought nearer to the light a point was reached at which the heliotropism disappeared but the shading reaction persisted. When moved still closer to the light they became negatively heliotropic but still without any change of the shading reaction. When moved still closer to the light, there was a short time when no shock movements could be obtained, but soon the illumination reaction appeared. At the same time the negative heliotropism became more prompt and precise. Finally, when the light was still further increased and allowed to act for a considerable time, even the illumination reactions frequently disappeared completely, and a most pronounced and compelling negative heliotropism held full sway. . . .
>
> It is very evident, then, that the invariable correlation of positive heliotropism with the shading reaction, which is required by Jennings's theory, does not exist. Both kinds of heliotropism may be associated with either the shading or the illumination reaction. Accordingly, it must be concluded that the heliotropic mechanism does not depend upon the mechanism for the shock movements, but that the two mechanisms are independent.[21]

The simplest method of determining whether or not the orientation of flagellates depends upon rapid changes in intensity of light or upon constant illumination can be furnished with the aid of intermittent light. We know that a striped muscle contracts only when a current is made or broken, but not while the constant current lasts. Hence a rapidly alternating current throws the muscle into tetanus, while the constant current has no effect. If it is the rapid change in the intensity of light which causes the swimming of a positively heliotropic *Euglena* to the light, an intermittent light, of a sufficient number of alternations per second, should be much more efficient than a constant light; while in case the positive heliotropism is determined by constant illumination, this should not be the case and the Bunsen-Roscoe law should hold.

Mast [348] has recently published experiments on the relative efficiency of the various parts of the spectrum by a method based on the assumption of the validity of the Bunsen-Roscoe law for the heliotropic orientation of these organisms. If his assumption [b] is correct, it contradicts the theory which Jennings and Mast have defended now for more than fifteen years; if his assumption is wrong, his experiments on the relative efficiency of various parts of the spectrum cannot be correct. Since, however, Mast's results with this method coincide with those by Loeb and Wasteneys [312] obtained by a direct method, it is very probable that the law of Bunsen and Roscoe holds for the heliotropic reactions of *Euglena* and unicellular flagellates in general, and, if this is true, the heliotropic reactions of unicellular algæ (*Euglena* included) are determined by light of constant intensity.

---

[b] He does not seem to have noticed that his method was based on this assumption.

CHAPTER XI

THE RELATIVE HELIOTROPIC EFFICIENCY OF LIGHT OF DIFFERENT WAVE LENGTHS

1. The validity of the Bunsen-Roscoe law for the heliotropic reactions of animals and plants leaves no doubt that these reactions are determined by the rate of photochemical processes. Heliotropic reactions depend, however, not only upon the intensity but also upon the wave length of light. Photochemistry shows that the most efficient wave length varies with the nature of the photochemical substance and that comparatively slight changes in the constitution of a molecule may bring about considerable changes in the relative efficiency of different wave lengths. The search for differences in the heliotropic effect of different wave lengths can be of service in detecting the nature of the photochemical substances responsible for heliotropic reactions.

The investigations on the relative heliotropic efficiency of different wave lengths have generally been undertaken for a different purpose, namely, to get information concerning the color sensations of animals. Graber gave it as the result of his observations that all animals which were fond of light were also fond of blue, and animals which were fond of dark were also fond of red.[180] He put animals into a box half of which was covered with transparent glass and half with an opaque object, and then counted the relative numbers of organisms in both halves of the box. He then replaced these screens by colored glasses and obtained the above-mentioned result. The writer showed that the animals are neither fond of blue nor of red but are oriented by the light in the same way as are plants, and the statement that animals which were "fond" of light also were "fond" of blue and those which were "fond" of "dark" were "fond" of red the writer explained in a simpler way, namely that the light filtered through red glass had a smaller orienting effect than the light filtered through blue glass.[287] Hence red glass acted like an opaque, blue glass like a transparent screen. This had already been known to be true for the heliotropic reactions of plants for which Sachs had shown that they occur behind a blue glass in the same way as behind common window glass, while behind red glass heliotropic reactions do not occur at all or occur very slowly as if the light were weak. The writer was able to show that the same is true for animals.[287] When positively heliotropic animals are put into a box covered with blue glass they go as rapidly to the window side as when the box is uncovered; while when it is covered with red glass the animals will go to the window but more slowly and irregularly. Behind a red screen they behave therefore as if they were exposed to weak light.

Blue glass is permeable not only for blue but also for rays which produce the sensation of green. Paul Bert [38] had already made experiments with positively heliotropic *Daphnia* in a solar spectrum and found that the animals "accouraient beaucoup plus rapidement au jaune ou au vert qu' à toute autre couleur."[a] Bert concluded from this that to the eye of a *Daphnia* those parts of the spectrum appear brightest which also appear brightest to the human eye. Bert's aim was to find out whether the sensa-

---

[a] This method of ascertaining the most efficient part of the spectrum is not reliable and has been replaced by other methods.

tions caused by light in lower animals are the same as those caused in a human being. But even if the relative efficiency of the various parts of the spectrum were the same for the sensations of brightness in human beings and for the heliotropic reactions in lower animals, it would not prove that the latter also have sensations of brightness. For we have no guarantee that the heliotropic reactions of lower animals are due to or accompanied by sensations of brightness. If the yellow-green rays are the most efficient in causing heliotropic reactions in an organism, it suggests only that in such an organism the photosensitive substance responsible for the heliotropic response is most easily decomposed by the yellow-green part of the spectrum.

A similar error of reasoning as that by Bert has recently been made by Hess. Hess corroborated what the writer had already pointed out, that the red rays of the visible solar spectrum are the least efficient, and he found, moreover, as Bert had found for *Daphnia,* that for the heliotropic reactions of a number of animals, from the fishes down, the yellow-green region of the solar spectrum is the most efficient. Now it happens that to a totally color blind human being, to whom the different parts of the solar spectrum appear only as shades of gray, the region $\lambda = 540\ \mu\mu$ in the yellow-green appears to be the brightest; while the red part of the spectrum gives a very faint sensation of brightness. From this similarity or apparent identity between the relative effects of different wave lengths upon the heliotropic effects of certain lower animals and upon the sensations of brightness of a totally color blind human being, Hess draws the conclusion that these animals are totally color blind. In our opinion the only conclusion which Hess has a right to draw is that the photosensitive substance which causes sensations of brightness in the eye of the color blind human being is either identical with or is affected in a similar way by light waves as is the substance giving rise to heliotropic reactions in certain animals. This assumption is entirely adequate and harmonizes better with the facts than the assumption made by Hess. The substance responsible for the sensations of brightness in the eyes of the totally color blind human being is visual purple which is bleached most rapidly by light of $\lambda = 540\ \mu\mu$. That our objection is justified is proved by the experiments of v. Frisch on bees.

v. Frisch[168] has shown by very ingenious and careful experiments that bees can be trained to discriminate between blue and yellow but not between different shades of gray. On a table were put square cardboards of different shades of gray and among them one blue piece of cardboard. On each gray square was put a watch crystal containing water, while the watch crystal on the blue contained sugar water. The bees of an observation hive visiting the sugar crystal were marked with a fine paint brush. After a sufficient period of training it was found that the marked bees always went directly to those crystals which were on a blue piece of cardboard, whether they contained sugar water or pure water; and when there was no sugar water on the blue cardboard they alighted on any blue object, *e.g.*, a blue pencil. The crystals and cardboard pieces were always renewed in different tests, to avoid any influence of odor. It was never possible to train the bees to select a piece of cardboard with a definite shade of gray among cardboards of different shades of gray.

Hess had shown that for the heliotropic reaction of

bees the yellowish-green part of the spectrum is most efficient, and he concluded that bees are totally color blind. To a totally color blind person the blue cardboard appears only like a shade of gray, and such a person is unable to learn to discriminate between a blue and gray piece of cardboard. v. Frisch's experiments support the conclusion that it is unjustifiable to use experiments on heliotropism to draw conclusions concerning light or color sensations. v. Frisch and Kupelwieser [169] have also demonstrated selective effects of different light waves for *Daphnia* which differ from those found for the eye of a totally color blind person, and their observations have been confirmed by Ewald.[147]

2. Hess's conclusions are in conflict with another group of facts. For many plants the blue region of the solar spectrum is the most efficient. Hess is, therefore, compelled to conclude that the heliotropism of such plants is different from that of animals, since it would seem preposterous to assume that swarmspores of plants should go to the light because they have sensations of color. He, therefore, assumes that plants are heliotropic—in the sense of the mechanistic theory—but that positively heliotropic animals go to the light on account of their love for brightness which is exactly the old viewpoint of Graber. It can be shown, however, that the difference in the heliotropism of animals and plants, which Hess assumes, is contrary to the facts, since there are heliotropic animals for which the blue rays are the most efficient, as for most plants; and there are green algæ for which the yellowish-green rays are most efficient, as for animals.

For many if not most plants the blue rays are the most efficient for inducing heliotropic curvatures. Blaauw [47] proved this in the following way: He exposed a row of

WAVE LENGTH

seedlings of *Avena* to a carbon arc spectrum for a certain time. The seedlings were then placed in the dark and after the proper time it was ascertained which part of the spectrum had induced heliotropic curvatures. By varying the duration of time of exposure to the spectrum it was found that with a minimal time of exposure only certain blue rays, namely, those of a wave length of 478 μμ caused heliotropic bending, while with longer exposure longer waves also became efficient. In this way the minimum duration of exposure for various parts of the spectrum was ascertained. Table IX gives his results.

TABLE IX

Duration of illumination, in seconds	Location of threshold in the spectrum, in micra
6.300	534 μμ
1,200	510 μμ
120	499 μμ
15	491 μμ
5	487 μμ
4	478 μμ
3	...
4	466 μμ
6	448 μμ

The red and yellow parts of the spectrum were ineffective for the intensity and time limits used and the optimum of efficiency was in the blue, in the region between 466 and 478 μμ.

A shorter series of experiments was made on the fruit bearers of *Phycomyces*, with the following results:

44 to 47 per cent. of the *Phycomyces* showed heliotropic curvatures

after 192 seconds of illumination at 615 μμ
after 192 seconds of illumination at 550 μμ
after 16 seconds of illumination at 495 μμ
after 32 seconds of illumination at 450 μμ
after 64 seconds of illumination at 420 μμ

The number of experiments was limited but they indicate an optimum between 495 and 450 µµ, in this respect agreeing with the results on *Avena*.

The fact then exists that for the heliotropic reactions of certain plants the blue rays are most efficient, while for the heliotropic reactions of a number of animals the yellowish-green rays are most efficient. But this statement cannot be generalized.

Loeb and Wasteneys determined the most efficient wave length of light for various lower organisms with the result that there are heliotropic animals for which the blue rays are as efficient as they are for plants; and that for different unicellular green organisms the optimum lies in different parts of the spectrum. They found, by a method similar to that used by Blaauw, that for the heliotropic curvature of the animal *Eudendrium* the most efficient part of the spectrum lies in the blue $\lambda =$ approximately 473 µµ.[311] The same was found by them for the larvæ of the marine worm *Arenicola*.

On the other hand, on investigation of two closely related forms of green flagellates, *Euglena* and *Chlamydomonas*, it was found [311] that they behave differently. For *Euglena viridis* the blue rays $\lambda =$ 470 to 480 µµ are especially efficient, while for *Chlamydomonas pisiformis* the most efficient part was in the region of $\lambda =$ 534 µµ, in the yellowish-green.[b] For another green algæ, *Pandorina*, Loeb and Maxwell had already found the greatest efficiency in the greenish-yellow.

---

[b] This would lead us, on the basis of the reasoning of Hess, to the conclusion that the unicellular plant *Chlamydomonas* has sensations of brightness, suffers from total color blindness (although it has no eyes), that it is not heliotropic, and that it is an animal: while its unicellular cousin, *Euglena*, has a highly developed color sense, has no sensations of brightness, is heliotropic, and is a plant.

The method used for these experiments by Loeb and Wasteneys is as follows:

A carbon arc spectrum, about from 18 to 23 cm. wide, was thrown on a black screen *SS* (see Fig. 34) with two slits *a* and *b* in the two different parts of the spectrum which were to be compared in regard to their heliotropic

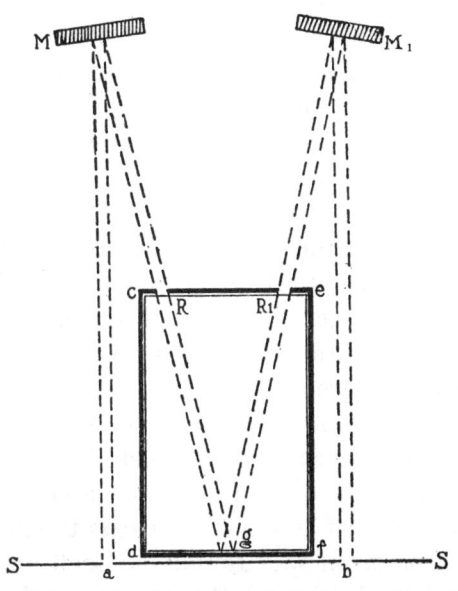

Fig. 34.—Method of determining the relative heliotropic efficiency of two different parts of the spectrum. (After Loeb and Wasteneys.)

efficiency. The two beams of light passing through the slits are reflected by the two mirrors *M* and M_1 into the square glass trough *cdfe* in such a way as to strike the same region *g* of the back wall of the trough. The glass trough is surrounded by black paper except at *R* and R_1, where the two beams of light enter from the mirrors. Before the experiment begins, all the organisms are collected in the spot *g* with the aid of an incandescent lamp. As soon as the spectrum is turned on, these organisms are simultaneously exposed to two different beams of light which come from the two mirrors M and M_1. When one type of light, *e.g.*, that from M, is much more efficient than the other coming from M_1, practically all the organisms are oriented by the light from M and move toward this mirror, collecting in the region R. When the relative efficiency of the two types of light is almost equal, the organisms move in almost equal numbers to R and R_1. By using as a standard of comparison the same region of the spectrum and successively altering the position of the other slit in the spectrum, we were able to ascertain with accuracy the relative efficiency of the different parts of the spectrum for the two forms of organisms. When the two parts of the spectrum which are to be compared are very close to each other, it is necessary to deflect the beams with the aid of deflecting prisms, before they reach the two mirrors.[311]

Experiments on the newly hatched larvæ of *Arenicola*, a marine worm, showed that the most efficient part of the spectrum was in the bluish-green of about $\lambda = 495$ $\mu\mu$, while for the larvæ of *Balanus eburneus* the most efficient part of the spectrum was found by Loeb and Maxwell, by Hess, and by Loeb and Wasteneys in the region of yellow and yellowish-green.[311]

Mast [348] made similar experiments on these organisms with a method in which the organisms were exposed to two beams of light of different wave length crossing each other at right angles. One light was kept constant while the other was made intermittent by a disk with a sector cut out rotating in front of the light. The size of the sector was varied until the organisms moved at an angle of 45° to the two beams. When this happened the heliotropic

TABLE X[1]

Relative Stimulating Efficiency of Different Regions in the Spectrum; Reduced to the Same Magnitude and Averaged so that They can be Readily Compared. Maximum in Heavy Type

Wave length in μμ		Euglena viridis		Phacus triqueter	Trachelomonas euchlora	Gonium	Arenicola larvæ	Lumbricus	Chlamydomonas	Blowfly	Pandorina	Spondylomorum
		Negative	Positive	Negative	Positive	Positive	Positive	Negative	Positive	Negative	Positive	Positive
Blue	422.4	2.50								0.24		5.62
	432.6	4.09								2.65		7.62
	442.8	10.41	4.29	7.22						(?)		7.62
	452.9	13.24	7.69	12.21	3.55	11.90	5.49	6.26		7.50	1.32	10.85
	463.1	14.86	12.70	16.49	15.81	16.60	11.84	9.72		9.46	3.57	13.87
	473.2	18.22	16.66	19.02	18.59	20.10	14.01	20.14	14.83	16.17	4.89	16.35
	483.4	20.40	18.30	20.07	20.03	17.50	16.63			18.91	7.86	
			20.86				20.24				9.15	
Green	493.6	14.46	18.30	16.47	17.64	19.45	17.34	18.23	17.96	19.39	12.51	16.60
	503.7	8.23	9.46	8.42	10.21	13.37	11.37	7.84	19.88	20.38	15.15	19.30
	513.8	3.06	3.13	1.98	5.16	5.76	6.58	2.34	15.70	19.50	18.78	19.62
	524.0	1.08	1.00	0.30?	1.05	2.71	2.67		8.01	16.21	20.10	19.27
	534.1					0.29	7.99			13.33	18.51	20.17
	544.3									8.41	17.67	18.20
	554.4									5.71	14.25	12.95
Yellow	564.5									3.42	9.42	7.00
	574.6									1.72	6.57	5.00
	584.8									1.05	3.39	2.95
Red	594.9									0.51	1.53	2.30
	605.0									0.40	0.60	1.70
	615.2										0.22	1.35
	625.3										0.09	1.00
	635.4									0.19		0.42
	645.6											

[1] Mast, S. O.: *J. Exper. Zool.*, 1917, xxii, 521.

efficiency of the two beams was considered equal. Mast's results, which are given in Table X, agree with those of Loeb and Wasteneys.

The error which Hess makes is of epistemological interest inasmuch as it shows the danger of false analogy. The real analogy for heliotropic reactions are forced movements and other tropisms, *e.g.*, galvanotropism or geotropism. Since forced movements (*e.g.*, in Ménière's disease) and galvanotropic reactions caused by a constant current through our head are not determined or accompanied by special sensations, the same may be true in regard to heliotropic reactions. This is not an idle assumption, since we know that the contraction of the iris of our eye under the influence of light is not accompanied by any sensation of brightness or color and such contractions occur also under the influence of light when the iris is excised. Hess ignores not only this analogy, but the whole existence of forced movements and of other tropisms, and he uses the color and light sensations of human beings, who are not heliotropic, to explain heliotropism in animals about whose sensations we know nothing. He fails to see that by this false analogy he dodges the real problem of heliotropism, namely, why the tension of symmetrical muscles changes upon one-sided illumination of an animal. For the explanation of this problem, we find assistance in the field of forced movements and of galvanotropism and of geotropism, but not in the behavior of totally color blind human individuals who show no trace of heliotropism.

The adoption of the false analogy between visual sensations and heliotropism makes it impossible for Hess to admit that bees should be heliotropic and at the same time be able to discriminate between blue and gray; while if we take cognizance of the analogy between heliotropism and the other tropisms we realize that the heliotropism of the bees and their reactions to blue are separate and independent phenomena, which need not be mutually exclusive and which in all probability depend upon different parts of the brain. When in certain cases the relative heliotropic efficiency of the various parts of the spectrum is identical with the curve for its apparent relative brightness to a totally color blind person, we may conclude that the photosensitive substances responsible for the two groups of phenomena behave similarly or may even be identical, but not that the sensations of brightness of the color blind and the heliotropic reactions of insects are identical or analogous phenomena.

Many mutants of *Drosophila* differ in regard to the pigments of the eye. It was natural to raise the question whether or not such hereditary variations of pigmentation of the eye influence the reaction of the flies to monochromatic light. McEwen investigated this possibility with the following result: " Colored lights which may be conveniently described as violet, green and red, are effective in the order named upon the insects whose eye color is lighter than the red eye of the wild fly. In the case of wild flies, and flies whose eyes are of a still darker shade called sepia, red is more effective than green'' (McEwen[549]).

CHAPTER XII

CHANGE IN THE SENSE OF HELIOTROPISM

We have stated that while in a positively heliotropic animal a one-sided illumination increases the tension of the muscles which turn the animal toward the source of light, in the negatively heliotropic animal the one-sided illumination must result in the opposite effect, namely, in a diminution of tension in the same muscles. As a consequence, the negatively heliotropic animal can turn more easily away from the light than toward the light.

Groom and Loeb [183] noticed that the larvæ of the barnacle upon hatching go directly to the light and gather at the light side of a dish, but that sooner or later their positive heliotropism may give way to an equally pronounced negative heliotropism. The stronger the light the more rapidly the larvæ are transformed into negatively heliotropic organisms. Later the reversibility of the sense of heliotropism was observed and studied in a number of organisms.[291] In a summary of the subject [300] (p. 470) the writer pointed out that this reversion was due either to a modification of photochemical processes or to an effect upon the nervous system. That an influence on the nervous system can indeed bring about a change in the sign of the reaction is very strikingly demonstrated in the following observation of A. R. Moore on starfish.[525]

Ordinarily, when a starfish which is moving in an aquarium is touched, it stops immediately and clings tenaciously to the surface of the vessel with its tube feet, so that it is impossible to remove the animal without injury to the tube feet. This normal response to sudden contact can be completely reversed by the administration of strychnine, so that when touched the animal loosens its hold on the bottom completely.

HELIOTROPIC TRANSFORMATION

The starfish poisoned with strychnine upon sudden touch withdraws all the tube feet, so that it can be moved about like an inert object. For this purpose 1 or 2 c.c. of a 0.5 per cent. solution of strychnine sulfate were injected into a starfish of medium size.

If the stretching out of the tube feet is due to an increase in the tone of the ring muscles (and a decrease in the tension of the longitudinal muscles) the drawing in is due to an increase in the tone of the longitudinal muscles of the tube feet. We therefore see that the same "stimulus," namely, a sudden touch, which causes one set of muscles to contract in a normal animal causes the antagonists of these muscles to contract in an animal poisoned with strychnine. We shall see that a number of cases of reversal of heliotropism may well find their explanation on this basis. On the other hand, the phenomena of solarization known in photography indicate that the sign of heliotropic response may also be changed by an excessive action of light on the photochemical substance. This effect, of course, may in the last analysis also result in an influence upon the central nervous system, such as that brought about by strychnine in Moore's experiment. We will now consider some cases more in detail.

The writer found [296] that certain fresh water crustaceans, namely Californian species of *Daphnia,* copepods, and *Gammarus* when indifferent to light can be made intensely positively heliotropic by adding some acid to the fresh water, especially the weak acid CO_2. When carbonated water (or beer) to the extent of about 5 or 10 c.c. is slowly and carefully added to 50 c.c. of fresh water containing these *Daphnia,* the animals will become intensely positive and will collect in a dense cluster on the window side of the dish. Stronger acids act in the same way but the animals are liable to die quickly. Esters, *e.g.,* ethylacetate, act also like acids and the addition of 1 c.c. of a grammolecular solution of ethylacetate to 50 c.c. fresh water also makes all the organisms positively heliotropic. Alcohols act in the same way. In the case of *Gammarus* the positive heliotropism lasts only a few seconds, while in *Daphnia* it lasts from 10 to 50 minutes and can be renewed by the further careful addition of some CO_2. The following table gives the minimal concentration of various acids and alcohols for the production of positive heliotropism in certain California species of fresh water copepods, and *Daphnia:*

	For Copepods	For *Daphnia*
Formic acid	0.006 N	
Acetic acid	0.006 N	
Propionic acid	0.005 N	
Butyric acid	0.004 N	
Valerianic acid	0.004 N	
Capronic acid	0.002 N	0.6 N
Ethyl alcohol	0.19 N	0.2 N
Propyl alcohol	0.054 N	0.05 to 0.1 N
Normal butyl alcohol	0.019 N	
Isobutyl alcohol		0.04 N
Amyl alcohol	0.011 N	

As far as alcohols are concerned each higher alcohol is about three times as efficient as the previous one, with the exception of amyl alcohol. This order of relative efficiency is also characteristic for the surface tension effects of these alcohols.[299]

It was of importance to find means of making these organisms negatively heliotropic. Moore[368] found that caffein makes the heliotropically indifferent fresh water crustacean *Diaptomus* intensely negatively heliotropic. It required the addition of 1.2 c.c. of a 1 per cent. solution of caffein to 50 c.c. of water to bring about this intense

HELIOTROPIC TRANSFORMATION

negativation. In two minutes all the animals are collected in a dense cluster on the negative side which lasts for about 35 minutes. A weak negative collection could also be obtained by adding 0.1 c.c. of a 0.5 per cent. solution of strychnine nitrate. Moore found that if the *Diaptomus* were first made positively phototropic by the addition of alcohol or acids, it was impossible to alter their response by the action of caffein, strychnine, or atropine. On the other hand, animals which had formed a negative collection under the influence of caffein if treated with carbonated water at once changed their response and swimming to the light side of the dish formed a positive gathering.

What causes these effects? The fact that alcohols make the organisms positively heliotropic suggested the possibility of a "narcotic" effect; the writer found, however, that narcosis requires a concentration of alcohols three times as high as the one required to produce positive heliotropism. He tried the effect of temperature on the reversal of the sign of heliotropism in *Daphnia* and found that lowering of the temperature enhanced the effect of acids in making the animals positive.[296]

The writer had found previously that in marine crustaceans and in larvæ of a marine annelid, *Polygordius,* the sense of heliotropism can be reversed by changes of temperature as well as by changes in the osmotic pressure of the sea water.[291] Increase in the osmotic pressure of sea water (by adding about 1 gm. of NaCl or its osmotic equivalent of other substances to 100 c.c. of sea water) made the negative animals positively heliotropic, and lowering of the concentration by adding 30 to 60 c.c. distilled water to 100 c.c. sea water made positive organisms negative. Negative larvæ of *Polygordius* or negative marine copepods could be made positive by lowering the temperature, and positive larvæ could be made negative by slowly raising the temperature. Since in the latter case the animals suffered from the high temperature the results were not so striking as in the case of the positivating effect of lowering the temperature. The same effect of the concentration of sea water and of temperatures was observed by Ewald for the larvæ of *Balanus perforatus*. He found, moreover, the interesting fact that a change of the ratio $\frac{Na}{Mg}$ in the sea water affected the sign of heliotropism of barnacle larvæ. An increase of Na made them more positive, an increase in Mg more negative.[144]

The larvæ of *Porthesia* are strongly positively heliotropic before they have eaten, while they lose their heliotropism almost completely after they have eaten.[287] The writer observed that male and female winged ants are strongly positively heliotropic but as soon as they lose their wings their heliotropism ceases.[287] McEwen[549] has found that when *Drosophila* is deprived of its wings its heliotropism ceases.

Holmes found that terrestrial amphipods are positively, while the aquatic amphipods are negatively heliotropic. By putting a terrestrial amphipod into water it became negatively heliotropic.[225]

That a reversal in the sense of heliotropism may be due to a nervous effect is suggested by an observation by Miss Towle[485] that a certain ostracod, *Cypridopsis*, can be made positively heliotropic by mechanical shock, and the writer noticed that indifferent fresh water *Gammarus* can be made negatively heliotropic by shaking them. In both cases the heliotropism lasts only a short time.

The attempt to explain all these reversals on the assumption of a change in the central nervous system meets with the difficulty that such reversals occur also in unicellular organisms which have no central nervous system. Thus the writer observed that *Volvox,* which occurred in the same ponds in California from where *Daphnia* came, could also be made positive by CO_2.[296] In swarmspores of algæ reversals of heliotropism are a common phenomenon. While these unicellular organisms have no central nervous system they may have synapses such as exist between different neura of metazoa. The writer is not sufficiently familiar with the behavior of synapses in higher animals to suggest that this condition is responsible for the changes in the sense of heliotropism.

We may finally discuss briefly a possible solarization effect. The writer found that it is possible to make animals generally negatively heliotropic with the aid of ultraviolet light.[296] If once rendered negative such animals will be negative not only to ultraviolet rays but also to the light of an incandescent lamp. A. R. Moore[366] found that the ultraviolet rays having such an effect have a wave length shorter than 3341 Å.U. Oltmanns had observed that *Phycomyces* is positively heliotropic in weak light, indifferent in somewhat stronger light, and negatively heliotropic in still stronger light. Blaauw found that when the illumination was strong the seedlings of *Avena* became negatively heliotropic.[47] He suggests the analogy with solarization effects in photography. The discovery of photodynamic effects by v. Tappeiner[477] adds to the possibilities which should be considered in this connection.

While *Drosophila* is usually positively heliotropic, McEwen has recently described a mutant of this species which is not heliotropic. This lack of heliotropic response is linked with a peculiar color—"tan"—by which the mutant is characterized. The character "tan" is sex linked. The daughters inherit the factor for the character from their fathers but do not show the character, while the sons inherit the factor from their mothers and do show the character. The lack of heliotropic reaction in this mutant is apparently not due to any structural defect in the eye (McEwen [549]).

Keeping successive generations of flies in the dark does not influence their heliotropism. F. Payne [550, 551] raised sixty-nine successive generations of *Drosophila* in the dark, but the reaction of the insects to light (as well as their eyes) remained entirely normal.

CHAPTER XIII

GEOTROPISM

1. When the stem of certain plants is placed in a horizontal position, the apex grows vertically upward and the root downward. The downward growth of the root is called positive, the upward growth of the apex negative geotropism. The writer has observed a similar phenomenon in a hydroid, *Antennularia antennina* [294, 300] and his observations were confirmed by Miss Stevens.[553] Animals as well as plants, therefore, show the phenomenon of geotropism.

These phenomena have given rise to a strange discussion, namely: What constitutes the "stimulus" in the case of geotropism? When a galvanic current is sent through a motor nerve the muscle answers with a contraction only when the current is made or broken, but not while a constant current is flowing through the nerve. The older physiologists were not able to form a mental picture of what happened in this case, and they cut the knot by invoking a verbalism, namely by calling the making or breaking of a current a "stimulus." This perhaps innocent verbalism then led to the less harmless dogma that only a rapid change could act as a "stimulus." Thus Jennings[253] and Mast[346] took it for granted that phenomena of orientation by light could only be produced by rapid changes in the intensity of light and not by constant illumination, since they had the *a priori* conviction that only a rapid change in the intensity of a galvanic current or of light is a "stimulus." The same difficulty arose in regard to the action of gravity upon orientation, since it was contrary to the definition of a "stimulus" that the mere permanent lying in a horizontal position should cause the apex of a stem to bend upward.

All these difficulties disappear if we take the law of chemical mass action into consideration. Light acts not as a "stimulus" but acts by increasing the mass of certain chemical compounds, and it is the mass of these products which is responsible for the effect of light. Now, mass action is not proportional to the rapidity of the *change* of acting masses but to the acting mass itself. When two sides of an organism are struck by light of different intensity the quantity of photochemical products on both sides becomes unequal. In galvanotropism the galvanic current alters the distribution of the mass of certain ions along the nerve elements.

It can be shown that gravitation acts by influencing the distribution of chemical substances in an organism. When the stem of a plant is put into a horizontal position certain chemical substances gather in greater concentration on the lower side of the stem; and this causes a difference in the velocity of chemical reactions between the lower and the upper side. As a result of this we notice the bending. In the normal upright position of the plant the same substances were distributed equally about the axis of symmetry.

The following facts may be offered as a proof for this statement.[526] When we put a piece of the stem of *Bryophyllum calycinum* in a horizontal position it soon bends and gradually assumes the form of a U with the concave side above (Fig. 35). This bending is due to the fact that the cortex on the under side of the stem grows in length while the cortex on the upper side remains unaltered

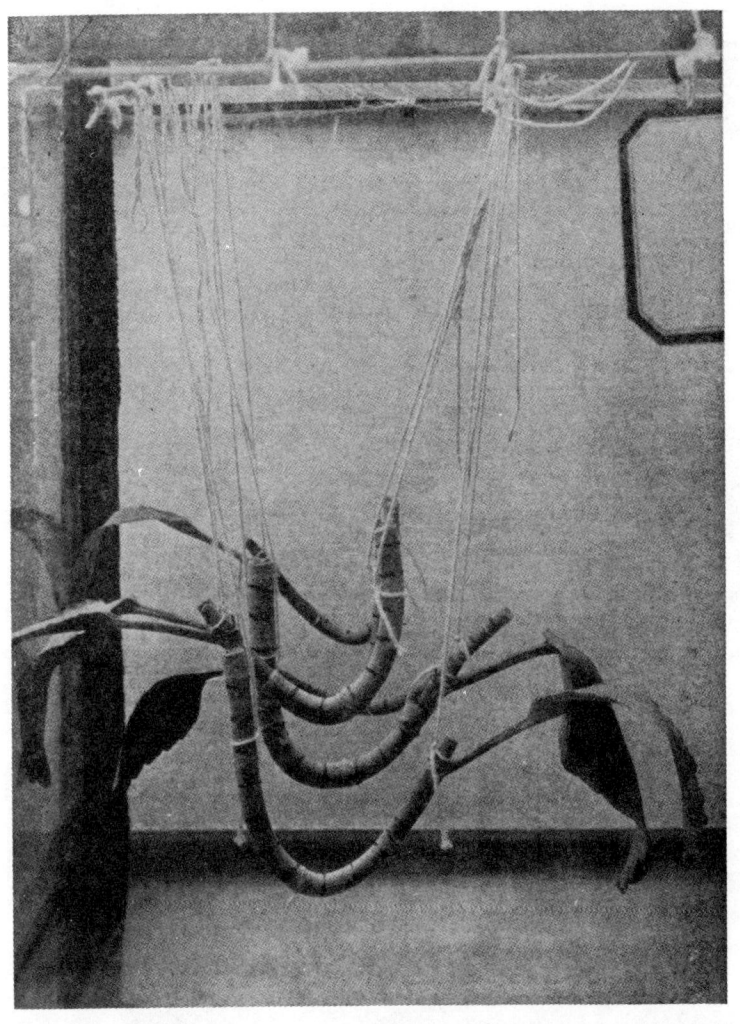

FIG. 35.—Geotropic curvature of stems of *Bryophyllum calycinum*. These stems were originally straight and suspended in a horizontal position. In about ten days they bent, becoming concave on the upper side. The black rings, made with india ink, which were originally parallel, remain unaltered on the upper side of the stems, while their distance increases on the lower side, indicating that the curvature is due to an increase in growth on the lower side (of the cortex) of the stem.

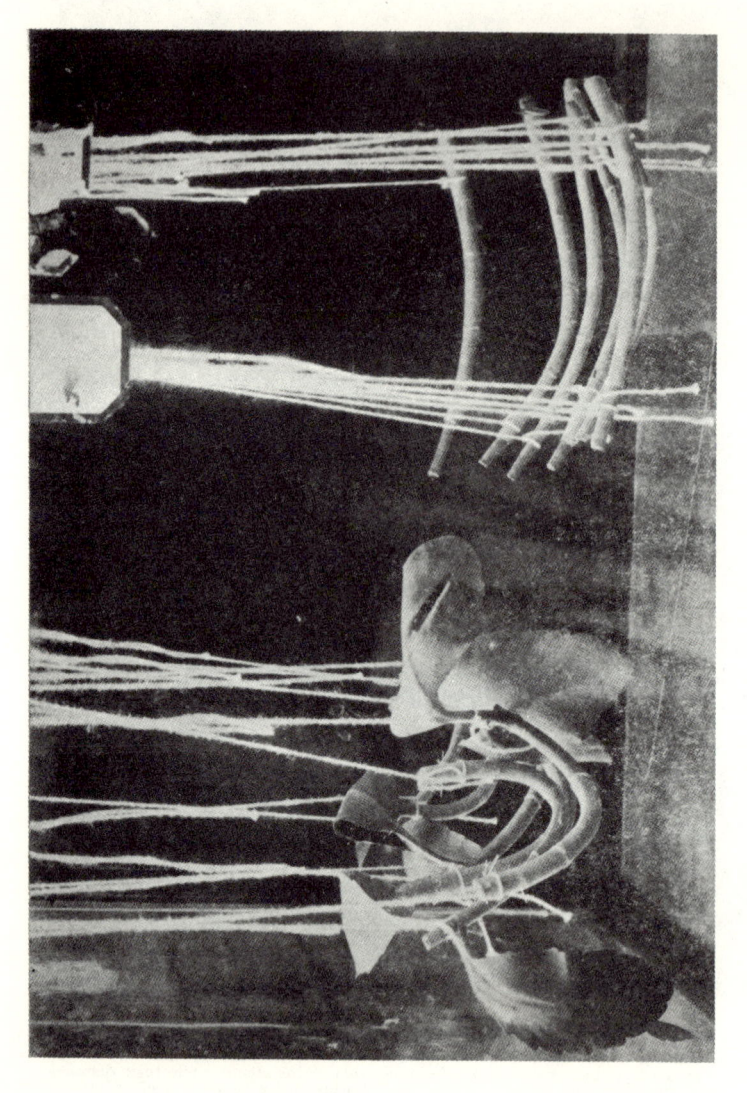

FIG. 36.—All stems were originally straight and suspended horizontally. The stems to the right, without leaves, bend much less in the same time than the stems to the left, with two apical leaves; the leaves supplying material for growth to the cortex of the stem on which the bending depends. This material collects in greater masses on the lower side of the cortex than on the upper side.

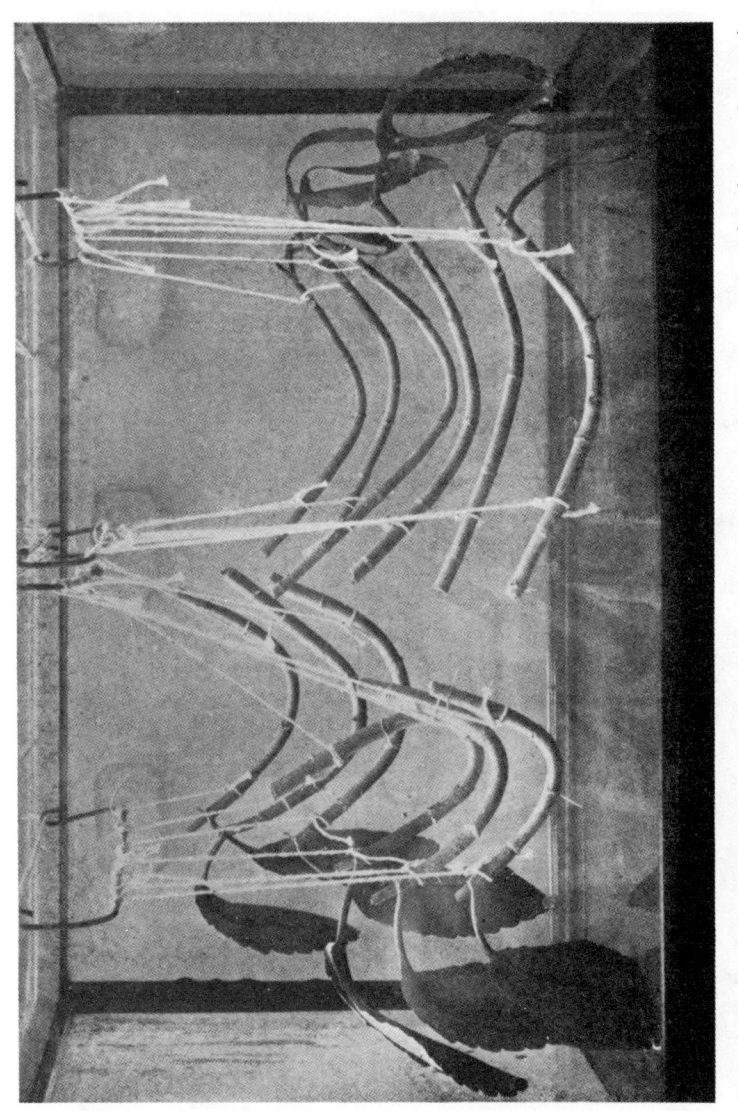

FIG. 37.—When the size of the leaf is reduced by cutting out pieces from the middle (stems to the right), the rate of geotropic bending of the stem is diminished, since the material sent by the leaf into the stem diminishes with the mass of the leaf.

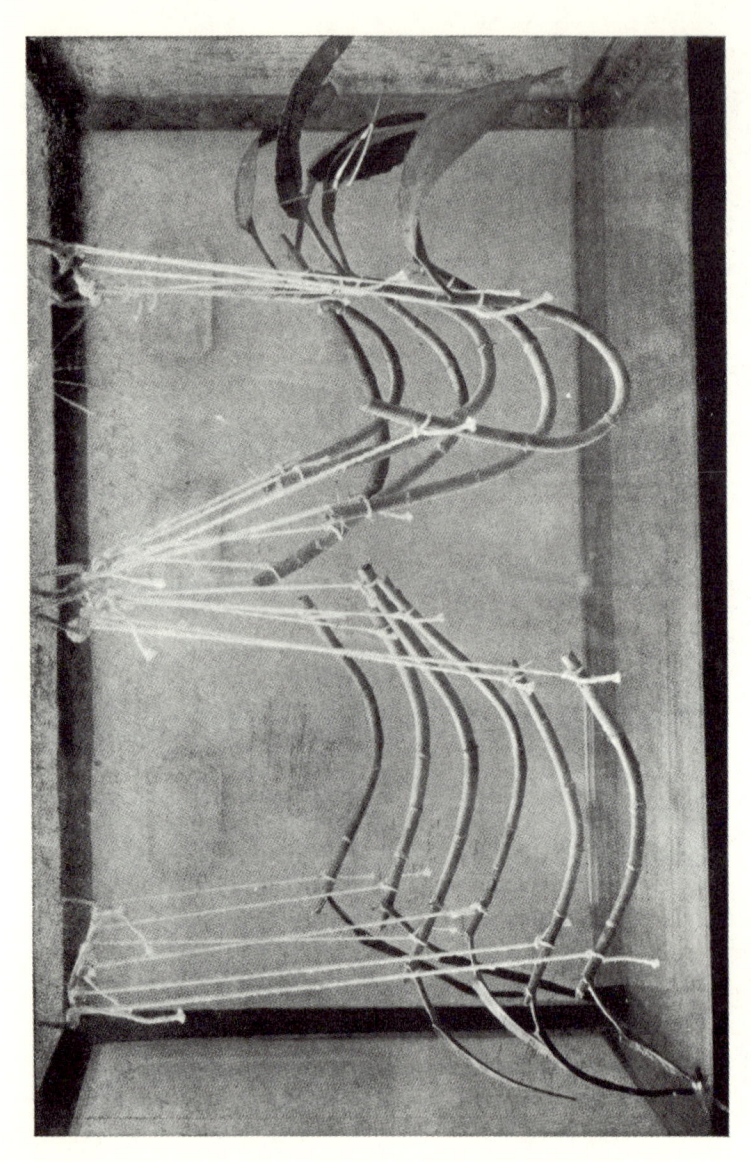

FIG. 38.—In stems to the left the size of the leaves is reduced by cutting off the lateral parts of the leaves. Stems to the right possess intact leaves. Notice that the geotropic curvature in the latter is much stronger than in the former.

(Loeb [542]). This can be demonstrated if we mark the cortex in definite intervals with india ink at the beginning of the experiment (Fig. 35). After some time the distance between these marks will increase in a certain region of the under side, while it remains constant on the upper side, and this difference causes the bending. This positive increase in length of the under side can only happen through growth, and this growth of the cortex on the lower side of the stem takes place at the expense of material furnished constantly by the leaves which send it in the direction toward the basal end of the stem. When we compare the rate of geotropic bending of horizontal stems without leaves and with one or two leaves at the apex, we find that the bending in the latter is much more rapid (Fig. 36), owing to the greater mass of material supplied for the growth of the cortex, and the same is true, if we compare the rate of curvature of stems having a whole apical leaf attached with that of stems having an apical leaf whose mass has been reduced by cutting off parts of the leaf (Figs. 37 and 38). The writer has shown in other experiments that under equal conditions leaves produce material fit for growth in proportion to their mass. It is, therefore, a safe inference that the influence of the mass of an apical leaf upon the rate of geotropic bending is due to the mass of material it sends into the stem. This material has obviously a tendency to behave like a liquid—which it probably is—and to sink to the lower level. It is, therefore, useless to look for a "gravitational stimulus." [526, 544]

What has been demonstrated in this case explains probably also why the apex of many plants when put into a horizontal position grows upward, and why certain roots under similar conditions grow downward. It disposes also in all probability of the suggestion that the apex of a positively geotropic root has "brain functions." It is chemical mass action and not "brain functions" which are needed to produce the changes in growth underlying geotropic curvature.

2. As long as animals are in such a position that their plane of symmetry goes through the center of the earth, the position of their eyes and limbs is symmetrical in regard to their plane of symmetry. If, however, we incline the animal we can bring about forced movements and forced changes of position of the same nature as those caused by injury of one side of certain parts of the brain. Thus we have seen that if we cut the left side of the medulla oblongata in a shark, its two eyes are no longer in a symmetrical position but the left eye looks down and the right eye up, when the shark is kept in a normal position. The same change can be brought about in a normal shark by the influence of gravitation. When the shark is kept in a position with its right side inclined downward, the right eye is turned upward, the left eye downward. This has nothing to do with light or vision, since it occurs in the dark just as well as in an illuminated room. The abnormal position of the eyes lasts as long as the animal is kept in this abnormal position. The experiment shows that if the plane of symmetry is no longer vertical, forced positions of the eyes can be produced of the same nature as those produced by one-sided injury of certain parts of the brain.

Just as in the case of one-sided injury to the medulla oblongata the changes in the position of the eyes are accompanied by changes in the position of the pectoral fins, so also when we put a normal shark with one side downward or half downward.[289] If the right side of such

GEOTROPISM

a fish is down the right pectoral fin is turned more ventrally, the left fin is turned more dorsally. This means, the tension of the muscles causing the right fin to press down and the left fin to press up is increased. This is the mechanism by which the normal "equilibrium" or more correctly the normal geotropic orientation of the animal is maintained. If the animal should accidentally roll to one side in its normal movements, the tension of the muscles of the pectoral fins would automatically change in such a way as to restore the normal orientation of the animal, whereby the plane of symmetry becomes vertical again. This "maintenance of equilibrium" is therefore a case of automatic orientation by gravitation comparable to the automatic orientation by light.

Geotropic changes in the position of the eyes are not confined to fishes,[320] they can be demonstrated in a rabbit and in crustaceans as well.

In vertebrates the reactions leading to the maintenance of equilibrium are apparently produced in the ear, since they disappear if the acoustic nerves are cut. Moreover, those parts of the brain whose injury brings about such changes in the position of the eye and the fins are parts of the receiving fibers from the acoustic nerve.[290]

It seems that some change in the pressure upon the endings of the auditory nerve is responsible for the effects. There are fine grains of $CaCO_3$—the otoliths—in the ear of many species pressing on the underlying nerve endings. If we put the median plane of a fish at an angle of 45° with the vertical, the otoliths will no longer press down equally in both ears. The idea first suggested by Delage that it is the pressure of the otoliths upon the nerve endings which is responsible for these reactions receives some support by a well-known experiment by

Kreidl.[270] A crustacean, *Palæmon,* loses its otoliths in the process of moulting and the animal curiously enough replaces them by picking up small grains of sand and putting them into its ears. Kreidl kept such crustaceans in jars free from sand but containing fine particles of iron which the crustaceans after moulting put into their ears. He expected that a magnet would now influence the animals as powerfully as gravitation, and this was the case. When, *e.g.,* he brought a magnet from above and the right near the animal the latter turned to the left and downward. The animal, therefore, behaved as if changes of pressure of the otolith upon the nerve endings determined its geotropic orientation.

The theory meets with two difficulties which, however, are not insuperable. First, removal of all the otoliths does not interfere with the normal orientation of the animal. This might find its explanation in the fact that the eyes act as a substitute. Delage had shown that if the otocysts are removed in crustaceans or cephalopods the animals lose their normal orientation more easily when they swim about excitedly than do normal animals. In order to show the effects clearly, however, it was necessary to blind the animals. Animals which were merely blinded but had their otocysts did not show these disturbances of equilibrium.[119]

The second difficulty is the fact that animals which possess naturally no otoliths are yet able to show such geotropic reactions, *e.g.,* certain crustaceans like *Gelasimus and Platyonichus.* We may assume that the pressure of liquids on the nerve endings may have a similar effect as the pressure of the otoliths.

The next question is, How does the pressure on a nerve ending bring about changes in the tension of muscles?

We suspect that this occurs through a change in mass action in the nerve endings, in analogy to our experiments on the influence of the mass of the leaf on the geotropic curvature of *Bryophyllum,* but experimental data are lacking.

3. We observe phenomena of geotropism in animals which have no ears, but this need not surprise us in view of the observations on geotropism in plants, and in hydroids (*Antennularia antennina*). The writer [289] had found that a holothurian (*Cucumaria cucumis*) has a tendency to creep upward when put on a vertical object until it reaches the highest level, where it remains. When put on a vertical plate of glass or slate, these animals creep untiringly upward if only the plate is turned 180° around a horizontal axis as soon as they have reached the highest point. It could be shown that light and oxygen supply have nothing to do with the phenomenon. Jennings observed that *Paramæcia* always gather at the highest point of a vertical tube and that they assume this position by active ciliary motion. Lyon [323] assumes that the body of *Paramæcia* contains substances of different density whose location is changed by changes in orientation of the organism to the center of the earth and that these changes automatically turn the animal again so that its oral pole is directed upward. It will then continue to swim in this direction.

4. It is known since Knight's experiments that centrifugal force can act like gravitation and we must assume that the centrifugal force leads to an alteration in the distribution of the sap or of other substances in the cell. This leads to differences in the rate of chemical reactions and may account for the phenomena of orientation under the influence of centrifugal force.

When an animal, *e.g.*, a shark or a pigeon, is rotated on a turntable, *during* rotation a nystagmus is observed in the motions of the eyes and sometimes also of the head. If the rotation is not too rapid the eyes move slowly in the same plane but in an opposite direction from the rotation of the turntable, until they form a maximum angle with their normal position in the head; then they rapidly swing back and the whole phenomenon is repeated. This phenomenon is called nystagmus. It depends upon the nerve endings in the semicircular canals, but is not dependent upon the motion or pressure of the lymph in the canals,[290, 319, 320] since the cutting out of the canals in the shark or the plugging up of the canals in the pigeon [141] leaves the phenomenon unaltered. When after some rotation the motion of the turntable suddenly stops, a nystagmus of the eyes or head in the same plane but in the opposite direction as during the rotation is observed.

Maxwell [554] has shown that if *Phrynosoma* is rotated on a horizontal plane with constant velocity and the eyes of the animal are closed, compensatory motions of the head are produced as soon as the angular velocity exceeds a certain value which was 8 seconds for a rotation through an angle of 45°.

CHAPTER XIV

FORCED MOVEMENTS CAUSED BY MOVING RETINA IMAGES: RHEOTROPISM: ANEMOTROPISM

THE experiments on forced movements show that we have three groups of forced movements, (1) right to left and left to right (circus movements); (2) forward movement, and (3) backward movement. The latter is not always possible. A fourth group, the rolling motions around the longitudinal axis may be omitted here in order to simplify the discussion.

The forced movements, called forth by the galvanic current, supported the idea that the nervous elements determining these motions must have a definite orientation and that this orientation bears some simple relation to the direction of motion caused by their activity. The experiments on the effect of blackening different parts of the eye indicate that the different parts of the retinæ of positively heliotropic insects are connected in a simple way with the main centers of the three types of forced movements: namely, the left eye is connected with the brain center causing motions from right to left (and the right eye with the center for the opposite motion); the lower halves of the retina with the forward movements, the upper halves with the backward movements.

We know through the work of Ewald Hering that each illuminated element on the human retina determines a definite motion of the two eyes which move as if they were a single organ, and that this motion is a function of the location of the illuminated element in the retina. This fact induced the writer to suggest in his first publication on tropisms that the act of focussing in our vision was simply a phenomenon of heliotropism. "The general principle of orientation of organisms to light is also manifested in our act of binocular vision which results automatically in such an orientation of the two retinæ that the image of the luminous point falls upon the two foveæ centrales of the retinæ" (which are symmetrical elements). In other words, when an object causes us to turn our eyes to it we are dealing with a phenomenon of forced (heliotropic) movement. In order to prove this it is necessary to show that a moving retina image can produce forced movements determined by the direction of motion of the luminous object. The difficulties inherent in the proof for such a statement lie in the general prejudice that the motions of an animal are directed to a purpose and it is, therefore, necessary to devise experiments which exclude the assumption of an interest on the part of the animal in the motion.

The writer observed years ago that when a fly is put on a rotating disk it rotates in the opposite direction from the disk. When the motion of the turntable ceases these compensatory motions of the fly stop also and none of the after effects mentioned at the end of the previous chapter are noticed.[286] This suggested that the so-called compensatory motions of insects on the turntable have a different origin from that of vertebrates. The phenomenon was explained by Rádl, who proved that the compensatory motions of insects on the turntable are produced in the eye and that they are due to the fact that the eye tries automatically to fix the same object.[447] This agrees with the observation of Lyon who had already

RHEOTROPISM

demonstrated previously that the compensatory motions of insects on a turntable stop when their eyes are blackened.[319] Such forced motions, due to the influence of the motion of the retina image, can be demonstrated in the Californian lizard *Phrynosoma blainvilli,* which is an ideal object for such experiments [298] and in this animal it is possible to separate these effects from the compensatory motions caused by centrifugal force.

It was accidentally observed by the writer that when the lizard *Phrynosoma* is kept at the window of a moving train with its eyes toward the window, a nystagmus of the head of the lizard ensues, the head moving slowly in a direction opposite to that of the moving train, as if to keep its eyes fixed on the objects outside—telegraph poles and trees, etc. The head moves until it is bent maximally, when it is brought back into its normal position with a quick jerky movement, and then follows again the apparent motion of the objects outside, and so on. These nystactic motions last for hours, in fact as long as the animal is kept with its head toward the window. As soon as it is turned around so that it cannot see the objects outside, the nystactic motions of the head cease. When the animal is put on a turntable and rotated slowly, vigorous compensatory movements can also be observed during rotation. If, however, the eyes of the lizard are closed during rotation these movements are considerably diminished though they do not cease entirely. They are also considerably diminished when the animal with its eyes open is rotated on a turntable surrounded by a high gray cylinder of cardboard which excludes the possibility of images of outside objects moving on the retina. We can also produce compensatory motions of the head if the animal is kept quiet and objects are moved in front of it, the eyes following the moving object.

It is of interest to separate the nystagmus or compensatory motions of eyes and head caused by the orienting effect of a moving retina image from those caused by the orienting effect of centrifugal force and this can be done easily in *Phrynosoma*.

When the lizard is rotated very *slowly* on a turntable with *its eyes closed, only very slight compensatory motions of the head and body are observed during rotation,* while very powerful compensatory motions are produced when the motion of the turntable is suddenly interrupted after a rotation lasting about thirty seconds.

When, however, the same experiment is made with *the eyes of the lizard open* the reverse is observed. The compensatory motions of the animal during rotation are exceedingly vigorous, while the compensatory motions of the animal after the interruption of the rotation are slight.

When the eyes of the animal are closed we are dealing only with the geotropic effect of passive rotation; when the eyes are open the orienting influence of the moving retina image is added algebraically to the orienting effect of centrifugal force upon the ear. These two influences act in the same sense *during* rotation and therefore are additive; while *after* the rotation they act in the opposite sense to each other. When we rotate the body of an animal passively to the right, *during* rotation the objects have an apparent motion to the left and the eyes and head of the animal are compelled to follow these moving objects, *i.e.*, to the left. The geotropic effect of passive rotation of the animal to the right also causes a motion of the eyes and head to the left and hence both effects are additive.

RHEOTROPISM

When a human being has been rotated passively to the right for some time, at the interruption of the passive motion the eyes move slowly to the right and return rapidly to the left. Only the slow motions give rise to the sensation of an apparent motion of the objects and hence after the sudden stopping of a passive rotation to the right the objects seem to such a person to move to the left. The geotropic after effect, after passive rotation to the right, consists in inducing passive compensatory motions to the right, *i.e.*, in the opposite sense of the orientation caused by the apparent motion of the visual objects. Hence in the after effect the orienting effect of the retina image and the centrifugal effect weaken each other.

Lyon [321, 322, 326] has shown that the phenomena which were formerly described as rheotropism in fish are due to the orienting effect of moving retina images. The reader is familiar with the fact that many fish when in a lively current have a tendency to swim against the current. This phenomenon was believed to be due to the friction of the water. Lyon showed that fish orient themselves just as well when they are put into a closed glass bottle, which is dragged through the water, although in this case they are not under the influence of any friction from the current. When the bottle is not moved the fish swim in any direction inside the bottle. It is obviously the motion of the retina images of the objects on the bank of the brook which causes the "rheotropic" orientation of fish. When driven backward by the current or when dragged backward in a bottle through the water, the objects on the bank of the river seem to move in the opposite direction. The animal being compelled to keep the same object fixed, an apparent forward motion of the fixed object changes the muscles of the fins in such a sense as to cause the animal to follow the fixed object automatically.

When such rheotropic fishes were kept in an aquarium and a white sheet of paper with black stripes was moved constantly in front of the aquarium the fish oriented them-

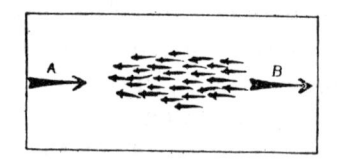

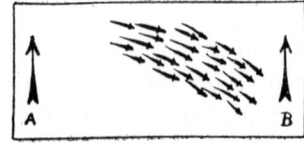

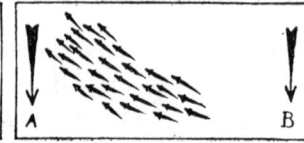

FIG. 39.—Influence of motion of the hand of an observer on the direction of the motion of a swarm of sticklebacks in an aquarium. The arrows indicate the direction in which the hand was moved. The swarm of fish moves always in the opposite direction in which the hand is moved. (After Garrey.)

selves against the direction in which the paper and its stripes moved. The phenomenon was more marked in young than in older specimens.

All the phenomena of rheotropism ceased in the dark or when the fish were blind.

Wheeler [508] has observed a phenomenon of anemotropism, namely that certain insects have a tendency to put the axis of their body in the direction of and against the wind. He considers this analogous to the phenomenon of rheotropism in fishes. The cause is also in all probability the tendency toward fixation of the moving retina image.

A very pretty demonstration of the orienting effect of moving retina images was discovered by Garrey in sticklebacks.[176] When a swarm of such fish was kept in an aquarium it was noticed that all the fish were oriented with the long axes parallel and that the whole school swam in a course parallel, but in a direction opposite, to that of the moving observer. If the observer remains stationary opposite the aquarium and moves an object, preferably white, which is held in the hand, the little fish at once respond by moving slowly and oppositely to that of the moving object. They can be thus made to move up or down or to the right or left (Fig. 39).

By experiments which space forbids us to report in detail Garrey has reached the conclusion that the motion of a near object causes an apparent motion of the whole horizon in the opposite direction and this apparent motion the fish tries to compensate by the motions of its body. This brings the observations on the stickleback into harmony with the general influence of moving retina images, consisting in a compensatory motion of the fish.

We have already referred to the fact that the influence of a moving retina image is capable of compensating the forced movement of a dog after a one-sided lesion of the cerebral hemispheres.

CHAPTER XV

STEREOTROPISM

Our orientation in space is determined by three groups of tropistic influences, two of which we have already discussed, light and gravitation. The third one is pressure on certain nerve endings of the skin. When the tactile influences on the skin of the soles of the feet are weakened (as is the case in locomotor ataxia), the patient finds it difficult to stand and walk in the dark. When he can use his eyes the difficulty is diminished, since the orienting effect of the retina image can compensate the tactile deficiency; just as we have seen that the effect of the loss of the ears in crustaceans can be compensated by the orienting influence of the eyes.

The rôle of tactile influences on the orientation of animals is most clearly demonstrable in starfish, flatworms, and many other animals, when put on their backs. The animals "right" themselves, *i.e.*, they turn around until the ventral surfaces or their feet are pressed against solid objects again. As the writer pointed out long ago,[293] gravitation has nothing to do with the phenomenon, since starfish will stick to solid surfaces with their tube feet even if by so doing their backs are permanently turned to the center of the earth. Unless the nerve endings on the sole of their tube feet are pressed against a solid surface the animals are restless and the arms move about until the feet are again in contact with solid bodies. This phenomenon of orientation the writer called stereotropism.

Quantitative investigations of this form of tropism are

STEREOTROPISM

still lacking and we must be satisfied with a few descriptive remarks.

Certain animals show a tendency to bring their body completely into contact with solid bodies, *e.g.*, by creeping into crevices. Without further experimental test this might appear as an expression of negative heliotropism, but it can be shown that this assumption would be wrong. *Amphipyra* is a positively heliotropic butterfly which, in spite of its positive heliotropism, shows the peculiarity that it creeps into crevices when given an opportunity. Such animals were kept in a box at the bottom of which was a square glass plate resting with its four corners on supports just high enough to allow the animals to creep under the glass plate. After some time every *Amphipyra* was found under the glass plate. This happened also when the glass plate was exposed to full sunshine, while the rest of the box was in the shade.[287]

The same stereotropism is found in female ants at the time of sexual maturity. When such animals are put into a box containing folded pieces of paper or of cloth, after some time every individual is found inside the folds. This happens also when the boxes are kept in the dark.[287]

The same form of stereotropism is found in many species of worms. When earthworms are kept in jars with vertical walls they are found creeping in the corners where their body is as much as possible in contact with solid bodies. It is this tropism which compels the animals to burrow into the ground.

Maxwell[349] kept *Nereis,* a form of marine worms, which burrows in sand, in a porcelain dish free from sand. Into the dish glass tubes were put, whose diameter was of the order of that of the worms. After 24 hours every tube was inhabited by a worm who made it its permanent abode. They even remained in the tube when exposed to sunlight which rapidly killed them.

We find the opposite, negative stereotropism, in many pelagic animals, *e.g.*, larvæ of the barnacle or of other crustaceans, which avoid contact with solids. The phenomenon is liable to interfere with heliotropic experiments.

The importance of stereotropism in animals was first pointed out by the experiments of Dewitz on the spermatozoa of the cockroach.[120, 121] He noticed that when a drop of salt solution containing the spermatozoa was put under a cover glass resting on low supports on a slide, the spermatozoa collect at the solid surfaces of the slide and cover glass, while the liquid between remains free from spermatozoa. When a small glass bead is put into the liquid the spermatozoa will also swim on the surface of the bead, never leaving it again. Dewitz is of the opinion that this stereotropism is of assistance in securing the entrance of a spermatozoon into the egg. The egg of the cockroach is rather large and the spermatozoon can enter it only through a micropyle. When the egg is laid it passes by the duct of the seminal pouch in which the female keeps the sperm after copulation. On passing the duct some spermatozoa reach the egg. Dewitz points out that these cannot leave the surface of the egg any more but are compelled to move incessantly on the surface of the egg until one of the spermatozoa by chance gets into the micropyle.

It is an important fact that different organs of the same organism react differently. We have already mentioned the tendency of starfish or flatworms to right themselves, *i.e.*, their ventral surface is positively their dorsal negatively stereotropic. The stolons of hydroids stick

STEREOTROPISM

to solid bodies, while the polyps bend and continue to grow away at right angles from solid bodies with which they come in contact. Thus the stem of *Tubularia mesembryanthemum,* a marine hydroid, grows in a straight line. When such stems, after their polyp is cut off, are put with one end in sand, the free end forms a new polyp and the stem continues to grow in a vertical direction upward. When, however, the stem is put near the glass wall as soon as the polyp grows out it bends away from

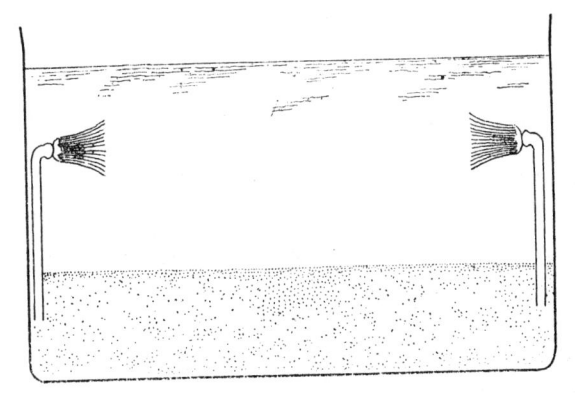

FIG. 40.—The regenerating polyp of *Tubularia* when in contact with the glass wall of an aquarium bends at right angles to the glass wall.

the solid wall, and the stem will now continue to grow at right angles to the vertical wall (Fig. 40).

This phenomenon raises the question whether or not the law of chemical mass action underlies phenomena of stereotropism. We have seen that this law dominates the phenomena of heliotropism, inasmuch as the Bunsen-Roscoe law is the expression of the influence of light on the mass of the photochemical reaction product. We have also been able to show that in the case of the geotropic curvature of *Bryophyllum* the mass of the apical leaf determines the rate of geotropical curvature of a horizontally placed stem. The only way in which the mass of the leaf could have such an influence is through the mass of substances it sends into the stem, so that this case of geotropism is a function of mass action. There are indications that the way contact with a solid influences the behavior of living matter is also through the influence on the rate of certain chemical reactions. The writer observed that the stolons of a hydroid, *Aglaophenia*, have a tendency to adhere to solid surfaces and not to leave them any more if they once reach them, and that as soon as such a stolon reaches a solid surface, *e.g.*, a piece of a glass slide, its growth is accelerated considerably. It was very astonishing to notice how much more rapid the growth of roots of *Aglaophenia* was when they were in contact with a solid body than when they grew in sea water. The rate of growth is the function of a chemical mass action (Loeb [543]).

CHAPTER XVI

CHEMOTROPISM

1. When we create a center of diffusion in water or in air we may theoretically expect orienting effects. Thus when a fine capillary tube containing a solution of a salt, *e.g.*, sodium malate, is put into a drop of water containing motile organisms, and the right side of an organism is turned to the source of diffusion, the diffusing molecules will collect in increasing concentration on that side. On the left side of the organism, no such increase in the concentration of molecules will occur. If now the molecules collecting on the right of the organism in increasing density are able to produce some chemical or some concentration chain effect, the two sides of the organism will be acted upon unequally and the tension of the symmetrical motile organs will no longer be the same. As a consequence the organism will turn until the mass of molecules or ions striking the organism in the unit of time will be the same for both sides. These effects only take place when the organism is close to the opening of the capillary tube, since the diffusion from the tube is slow.

It is obvious, however, that it is difficult to provide experimental conditions which give exact chemotropic reactions. First of all, if the diffusion is rapid the differences in concentration of the effective chemotropic substance on two sides of an organism are too slight to result in a turning movement. A second condition which is liable to vitiate the result are the unavoidable convection currents due to changes or differences of temperature. In order to get clear results a method must be used which prevents a rapid diffusion of the substance; and, moreover, the current of diffusion must be confined to an almost straight line. It is possible that Pfeffer's method satisfies this condition.[424, 425] He introduced the substance to be tested for its chemotropic effect into a capillary tube, the end of which was then sealed. The other end was pushed into a drop of water containing the suspension of the organisms whose chemotropism was under investigation. From this capillary the diffusion was extremely slow. Moreover, the current of diffusion was approximately linear at the orifice. Hence the test for the existence of positive chemotropism was perhaps possible. When an organism, struck sidewise by the line of diffusion near the opening of the capillary tube, turns toward the tube going into it, some probability of positive chemotropism exists; and when all the organisms coming near the orifice of the tube are thus compelled to go into it, the probability may become certainty, provided that the substance used does not paralyze the organism and therefore act as a trap, allowing the organisms to come in but not to go out. The capillary tubes used were of 10 to 15 mm. length and of a width of about 0.1 mm. Pfeffer and his pupils found that the spermatozoa of ferns go in large numbers into a capillary tube containing sodium malate in a concentration of 0.01 per cent. (a solution ten times as diluted is still slightly active). This effect of the malate is specific in this case and this indicates that either a definite chemical action of the malate ion or a specific permeability of the organism for it is the source of the chemotropism. Such specific chemotropic effects are not rare, since Pfeffer found that *Bacterium termo* and *Spirillum undula* are positively chemotropic

CHEMOTROPISM

to a liquid containing 0.001 per cent. of peptone or of meat extract. It is stated that cholera bacilli are strongly attracted by potato sap. Pfeffer found also that the spermatozoa of certain mosses are positively chemotropic to cane sugar solution in dilutions of 0.1 per cent.

Pfeffer's work preceded the discovery of electrolytic dissociation, and his pupils Buller[89] and Shibata[465] made some of the additions required by the theory, namely, that it is the malate anion which acts in the case of the spermatozoa of the ferns, and that when the anion is offered in the form of malic acid the H ion counteracts the effect of the malate anion.

Shibata made extensive experiments on the chemotropism of the spermatozoa of *Isoëtes*[465] which he found positively chemotropic for the malate anion, and also for the succinate, tartrate, and fumarate anion, when offered in the form of their neutral salts. The anion of the stereoisomere of fumaric acid, namely of maleïc acid, was without effect. This indicates a high degree of specificity of these reactions. Neutral sodium malate acted best in dilutions from m/100 to m/1000, but some action could still be discovered in m/20,000 solutions.

When malic acid was used no positive chemotropism could be discovered in solutions of m/100 or above on account of the contrary effect of the hydrogen ion, and the spermatozoa of *Isoëtes* did not even go into capillary tubes containing m/1000 malic acid. When any acid other than malic was added to sodium malate the motion of the spermatozoa into the tube was prevented, even a m/6000 HCl solution still had such an effect.

Shibata studied especially the mode by which the spermatozoa are oriented chemotropically by malates and found that the reaction consists always in a turning of the axis of the body of the spermatozoa toward the capillary tube containing malates or succinates, as the tropism theory demands.

When the capillary tube and the surrounding medium contain the same solute for which the organisms are positively chemotropic, they will not go into the tube unless the concentration in the tube is a *definite* multiple of the concentration of the outside solution. Thus Pfeffer found that the concentration of sodium malate in the capillary must be at least thirty times as great as in the outside solution to induce the spermatozoa of fern to move into it, and in the case of *Bacterium termo* the solution of meat extract in the tube had to be at least four times as great as the outside solution. In the case of *Isoëtes* spermatozoa Shibata found the ratio of about 400 to 1. This constancy of the ratio is known as Weber's law, which therefore holds for chemotropic phenomena.

Lidforss [281] found with the aid of Pfeffer's method that the spermatozoa of *Marchantia* are positively chemotropic to certain proteins, especially egg albumin, vitellin from the egg yolk, hemoglobin, and mucin of the submaxillary gland; blood albumin, casein, and legumin were less effective. The lowest concentration for hemoglobin solutions and for egg albumin was 0.001 per cent.!

It may also be stated that Lidforss found a chemotropic effect of proteins upon the direction of growth of pollen tubes.[280]

Bruchmann [81] found that the spermatozoa of *Lycopodium* were positively chemotropic to the watery extract in which pieces of the prothallium had been boiled. Pfeffer's capillary method was used. They showed also positive chemotropism to the citrate anion. Thus, sodium citrate was efficient in a 0.1 to 0.5 per cent. solution. The

CHEMOTROPISM

lower limit was a little above a 0.001 per cent. solution. The effect of the free citric acid was a mixed one since the spermatozoa were negative to H ions and positive to the citrate anion. Instead of being able to use a 0.1 per cent. solution, as in the case of the sodium salt, a 0.01 per cent. solution was the highest concentration to which they were positively chemotropic. This means that the hydrogen ion of citric acid solutions above m/1000 repel the spermatozoa, while when solutions of m/2000 or below are used the hydrogen ion effect no longer inhibits the positive effect of the citrate anion. In addition the validity of Weber's law could be demonstrated. The spermatozoa were indifferent to malates, oxalates, and many other salts, as well as to sugar and proteins.

2. While all the botanical observers, from Buller on, had found that the hydrogen ion has only a preventive effect upon the positive chemotropism of lower organisms, Jennings tried to show that acids have a positive effect, especially when in low concentrations.[250] But his concentrations are not quite as low as he seems to assume, since a 1/50 per cent. (m/180) HCl solution, toward which he believes to have proven positive chemotropism of *Paramœcia*, is a deadly concentration.[a] Jennings's interest in the problem was aroused by a phenomenon of aggregation, not infrequently found in the suspensions of infusorians.

It is well known that when certain infusoria are left undisturbed they do not remain scattered, but gather in more or less dense groups. Thus, if they are mounted on a slide in a thin layer of water, soon dense aggregations will be formed in certain areas, while the remainder of the

---

[a] The cells of the stomach resist a much higher concentration of HCl but this is an exception. Infusorians, fish, and organisms in general are killed in a short time in m/180 HCl or in a much lower concentration of acid. Thus *Fundulus* does not live more than one hour in m/3000 HCl or HNO_3. (Loeb, J., and Wasteneys, H., *Biochem Z.*, 1911, xxxiii, 489; 1912, xxxix, 167.)

slide will be nearly deserted. One of the first investigators to describe this phenomenon was Pfeffer. He observed its occurrence in *Glaucoma scintillans*, and less markedly in *Colpidium colpoda*, *Stylonychia mytilus*, and *Paramæcium*. Pfeffer was inclined to believe that these aggregations were due, partly at least, to a contact stimulus, resulting from a striking of the organisms against small solid bodies, and especially against each other.[250]

This conclusion of Pfeffer may after all be correct, since it has been shown that sea water containing jelly from the egg of a sea urchin causes spermatozoa to stick together for some time when they impinge upon each other. This agglutination no longer occurs when the spermatozoa are immobilized. Jennings came to the conclusion that these aggregations of infusorians are due to the fact that they can go into a weak concentration of acid, while they cannot escape from such a weak concentration; and since *Paramæcia* themselves produce CO_2 he assumed that the CO_2 produced by themselves acts as a center of attraction for other *Paramæcia*. In order to prove this he used the following method:

The organisms were studied in a thin layer of water, by mounting them on a slide covered with a large cover glass supported near its ends by slender glass rods. Their reactions were tested by introducing with a capillary pipette a drop of the substance in question beneath the cover glass, or in some cases by allowing it to diffuse inward from the side of the cover glass.[250]

Thus Jennings introduced a drop of 1/50 per cent. (m/180) HCl on a slide containing *Chilomonas*. Very soon a somewhat denser ring of these individuals was formed around the drop (Fig. 41). A 1/50 per cent. HCl solution paralyzes (and soon kills) *Chilomonas* or *Paramæcia* and hence the surface of the drop must act like a trap into which the organisms will steadily swim, without being able to swim back. This will naturally increase the density of organisms around the drop and may give rise to a ring formation around a high concentration of HCl although the organisms are not positive to the acid. Jennings found, however, that when such organisms are in a drop of weak acids which do not paralyze the organisms quickly, e.g., 1/50 per cent. acetic or in CO_2 solutions, they become negative to the surrounding neutral medium (H_2O or hay infusion) and stay in the acid. He, therefore, assumes that the organisms are positive to weak acid, and

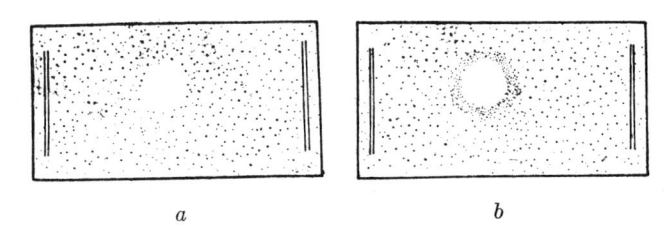

FIG. 41.—Reaction of *Chilomonas* to a drop of 1/50 per cent. HCl. *a*, preparation immediately after the introduction of the drop (no organisms either within or gathered about the drop). *b*, the same preparation a few minutes later. (After Jennings.)

negative to strong acid as well as to their natural neutral or faintly alkaline medium.

This negativity to their natural surroundings when in weak acid as well as to strong acid when in weak acid Jennings does not interpret in terms of the tropism theory, and in this he is probably correct. He interprets both phenomena as a trap action due to the asymmetry of certain infusorians; a sudden change in the concentration of a solution causes a reverse of the stroke of their cilia by which the organism is driven back. When the old normal stroke of the cilia is resumed the direction of the locomotion is changed on account of the asymmetrical arrangement of the cilia. This happens when the organisms go from weak into strong acid or from weak acid into a neutral medium. In this way a collection of the organisms at the surface of a drop of acid may be brought about. This phenomenon is not tropistic in the strict sense of the word, and as a matter of fact *Paramæcium* is not positively chemotropic to acid of any strength.

Barratt[24] investigated the chemotropism of *Paramæcia* for varying concentrations of different acids with

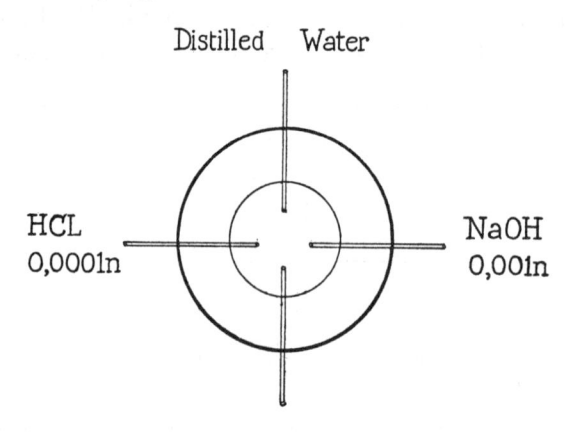

Distilled Water

HCL 0,0001n   NaOH 0,001n

Hay Infusion

FIG. 42.—Method of proving that *Paramæcia* are not positive to acids of low concentration. (After Barratt.)

Pfeffer's method of capillary tubes, counting the number of individuals going into the tube containing acid and comparing it with the number going simultaneously into a control tube containing only distilled water free from CO_2 (Fig. 42).[b] The acids used varied from 0.001 N to 0.0001 N. The results were unequivocal. Toward solutions of 0.001 N the *Paramæcia* are negative and possibly

---

[b] In addition two other controls accompanied the test, namely, one tube containing hay infusion (the natural medium of the organisms) and one alkali.

CHEMOTROPISM

also slightly negative to acids as weak as 0.0001 N. In no case, not even with the weakest acid, was it possible to prove the existence of positive chemotropism for acid (or base). The number of *Paramæcia* which went into a tube containing, *e.g.*, 0.00002 N acid, was on the average not greater than that which went into the control tubes. The tubes were sufficiently wide so that the *Paramæcia* could and did move into the tubes. Barratt, therefore, concludes that acids have only a repelling action upon *Paramæcia* which, however, diminishes or disappears when the hydrogen ion concentration approaches that of distilled water.

The observations of Barratt contradict the statement that *Paramæcia* are positive to weak acid. We have seen that when spermatozoa or swarmspores are positive to malates this can be elegantly shown by Barratt's method. The same method has shown that when even a trace of acid is added to the neutral malates this positivity disappears. By testing systematically all concentrations of different acids within the range to be considered, Barratt found no trace of any positivity to or any trap action by weak acid for *Paramæcia*. It may be true, however, that when the organisms are in very dilute acid neutral or faintly alkaline water repels them in the way described by Jennings.

Barratt states also that there is nothing to support Jennings's assertion that the CO_2 given off by the *Paramæcia* causes the aggregation in their natural medium, since they are not positive to low concentrations of hydrogen ions. The natural aggregations of infusorians may be due, as Pfeffer suggested, to transitory agglutinations when *Paramæcia* impinge upon each other, and the stickiness or tendency to agglutinate may possibly be increased by certain substances produced and excreted by the organisms themselves, *e.g.*, CO_2.

3. The results obtained with the spermatozoa of ferns and mosses by Pfeffer and other botanists led some authors to the tacit assumption that the spermatozoa of animals were positively chemotropic toward substances contained in or secreted by the eggs of the same species. Some accepted this assumption without test, others made tests which they considered adequate but which seem doubtful, and it may be of some interest to discuss the subject, since far-reaching conclusions might be based on these experiments. Pfeffer's method of testing for chemotropism with the aid of the capillary tube has proved satisfactory and the application of this method has shown that the spermatozoa of certain animals, *e.g.*, of sea urchins, are not chemotropic toward substances contained in or given off by the egg. Thus Buller, who had worked in Pfeffer's laboratory on the chemotropism of the spermatozoa of ferns, investigated carefully and extensively the question whether or not the spermatozoa of echinoderms are positively chemotropic for egg substances.[90] His results were entirely negative. Thoroughly washed, ripe unfertilized eggs of *Arbacia* (Naples) were put into a small volume of sea water for from 2 to 12 hours.

Capillary glass tubes, about 12 mm. long and 0.1 to 0.3 mm. internal diameter, and closed at one end, were then half filled with the (supernatant) sea water (which had contained the eggs) by means of an air pump. The tubes were then introduced into a large open drop of sea water, in which fresh, highly motile spermatozoa were swimming. If the eggs excrete an attracting substance it was argued that it should be present in the tubes, and the spermatozoa should collect there. . . . No attraction into the tube could be observed. Except for a surface-contact phenomenon to be further discussed, they went in and out with indifference. Apparently, therefore, the water which had contained the eggs exercised no directive stimulus on the spermatozoa whatever.

I then attempted to find some substance which could give a chemotactic stimulus to spermatozoa. The substances tested were such as are known to give a directive chemical stimulus to many protozoa, the spermatozoa of ferns, pollen-tubes, etc. The following solutions were tried by the capillary tube method: distilled water; meat extract 1 per cent.; KNO_3 10 per cent., 2 per cent.; NaCl 5.8, 2.9, 0.58 per cent.; K_2 malate 1, 0.1 per cent.; asparagin 1 per cent.; glycerine 5 per cent.; grape sugar 18, 9, 4.5, 2.25 per cent.; peptone 1 per cent.; alcohol 50, 25, 10 per cent.; diastase 1 per cent.; oxalic acid 0.9, 0.09, 0.009 per cent.; nitric acid 1, 0.1, 0.01 per cent.

No definite chemotactic reaction—neither attraction nor repulsion—was observed in any case. Into tubes containing the weaker solutions the spermatozoa went in and out with apparent indifference. . . . On coming into contact with strong acid solutions (oxalic acid 0.9, 0.09 per cent.; nitric acid 1, 0.1 per cent.) the spermatozoa *were killed, and thus formed slight collections.* They were thus not able to avoid acids by means of a negative chemotactic reaction.[90]

Other authors, *e.g.*, Dewitz and the writer, have also reached the conclusion that the egg of the sea urchin contains no substance for which the spermatozoon of the same species is positively chemotropic, and that Buller's conclusions that positive chemotropism plays no rôle in the entrance of the spermatozoon of sea urchins into the egg is correct.

F. Lillie has recently expressed the opposite view, namely that the egg of the sea urchin contains a substance to which the spermatozoa are positively chemotropic and to which he gave the name "fertilizin."[283] He first tried Pfeffer's correct method with capillary tubes with negative result, just as Buller and the rest of the observers. Instead of concluding that the spermatozoa are not chemotropic he discarded the method and used Jennings's method, stating that it gives "incomparably more delicate results than Pfeffer's method of using capillary tubes" (p. 533). Lillie found with this method that the spermatozoa of *Arbacia* are positively chemotropic to

H_2SO_4 of a concentration as high as N/10 and that they are never negatively chemotropic, not even to the highest concentrations of the strongest acid. It seems to the writer that Lillie's observations are more naturally explained on the assumption that when an acid is sufficiently strong and concentrated, *e.g.*, N/10 HNO_3 or H_2SO_4, it will paralyze and kill the spermatozoa, and that when a drop of such acid is introduced in sea water containing spermatozoa, a somewhat denser ring of the organisms will be formed around the surface of the drop on account of this action of the acid.

With the same method Lillie tried to prove that the spermatozoa of *Nereis* and *Arbacia* are positively chemotropic to extracts of their own eggs.[283] He proceeded as follows: A suspension of *Arbacia* sperm, freshly made, was put under a raised cover slip and a drop of the supernatant sea water which had been standing over eggs (as in Buller's experiments) was introduced under the cover slip. Observation with the naked eye showed that around this drop of egg-sea water immediately a dense ring of spermatozoa formed and behind this a clear external zone was formed about 1.2 to 2 mm. wide. The dense ring then broke up into small agglutinated masses. In Lillie's opinion the formation of this dense ring of spermatozoa at the periphery of the egg-sea water is the expression of a positive chemotropism of the spermatozoa for a substance contained in the egg-sea water, the "fertilizin." He assumes that the spermatozoa near the drop of egg-sea water all swim to the egg-sea water, leaving a clear space behind them. While this explanation of the ring formation might be true—if supported by a direct chemotropic method like Pfeffer's—it can be shown that the ring formation is in all probability due to an entirely different phenomenon which has no relation to chemotropism or any other tropism.

Buller had already observed that the supernatant sea water of sea urchins contains a substance which causes the agglutination of spermatozoa.[90]

A drop of sea water in which eggs had been deposited was placed upon a slide and a drop containing spermatozoa near it. On joining the drops a large number of small balls were formed in a very few seconds. When very numerous spermatozoa were present the balls became 0.1 mm. in diameter, containing many thousands of spermatozoa packed together in a dense mass.

Buller explains the phenomenon as being due to small bits of egg jelly floating in the sea water

so small that they will (like spermatozoa) pass through ordinary filter paper and, so transparent that one cannot directly see them. A few spermatozoa become attached to each piece of jelly, the presence of which may be inferred from the manner in which the small groups of spermatozoa move about. Owing to the length of the spermatozoon, although its head may be imbedded in a jelly particle, the tail may remain partly free. The little collections of spermatozoa thus move about hither and thither in no particular direction. When two such groups come by accident into contact they fuse. Certain of the spermatozoa adhere to both little masses of jelly and lock them together. The fused mass combines with other simple and fused masses, and so on.[c]

The writer was able to show that when the jelly of the egg of *Strongylocentrotus purpuratus* is dissolved by an acid treatment the eggs when washed and transferred to sea water no longer give off agglutinating substances, while the acid sea water containing the dissolved jelly, when rendered neutral through the addition of alkali, will cause the agglutination of sperm.[302] While all the jelly can be washed off with an acid treatment in the egg of *purpuratus,* the same is not true for the egg of *Arbacia*

---

[c] This explanation of the fusion of two clusters to a larger one is perhaps not correct. The writer is inclined to ascribe it to the adhesion or agglutination of the spermatozoa of two neighboring clusters with each other, due to a sticky surface on the sperm head.

of Woods Hole. Here the acid treatment does not as a rule dissolve all the jelly, or possibly some new jelly may be given off by the egg.

While Buller may be correct in assuming that microscopic pieces of the egg jelly form the center of these sperm clusters, the writer reached the conclusion that the dissolved mass of the jelly makes the surface of the spermatozoa transitorily sticky, so that if they impinge against each other they will stick together for some time, until the sticky compound formed by the jelly on the sperm head is dissolved by the sea water, which occurs after a short time.

This agglutinating effect of the egg-sea water upon the sperm of *Arbacia* gives rise to that ring formation which Lillie considers a proof of positive chemotropism. When a drop of egg-sea water is put into a sufficiently dense suspension of spermatozoa, the spermatozoa at the surface of the drop will agglutinate into practically one dense ring around it, and through the diffusion of some of the dissolved jelly through this ring numerous little clusters will form at the external periphery of the ring, and these clusters will fuse with the ring. In this way the clear region behind the ring originates. The process of fusion continues inside the ring with the result that the latter breaks up into numerous bead-like spherical clusters as Lillie described. In a former paper the writer has pointed out the analogy between the phenomena of transitory sperm agglutination (under the influence of egg-sea water) and surface tension phenomena, inasmuch as two small clusters upon coming in contact fuse into one larger one and inasmuch as elongated clusters break up into two or more spherical clusters.

The ring formation described by Lillie has, therefore,

in the opinion of the writer no connection with positive chemotropism.[d]

4. The method of Pfeffer cannot well be used for larger organisms. Barrows [25] has devised an apparatus which allowed him to test quantitatively the chemotropic reactions of *Drosophila*. The flies which are positively heliotropic were allowed to go to the light inside of a narrow hollow groove. At a certain spot of the groove two glass bottles were inserted with their openings opposite each other, one of which contained the substance to be tested for chemotropic efficiency, while the other served as a control. The number of flies which on their path were deviated by the bottle containing the substance to be tested were counted and their number compared with that going into the control bottle. The collection of odorous matter in the groove was removed by suction. In this way it was possible to ascertain that the flies are positively chemotropic to ethyl and amyl alcohol, acetic and lactic acid, and to ether. The chemotropic effect of alcohol was increased through the admixture of traces of an ester, *e.g.*, methyl acetate.

In describing the manner of reaction of these flies, Barrows makes the statement that when the odor is weak the fruit fly "attempts first to find the food by the method of trial and error, but as the fly passes into an area of greater stimulation, these movements give way to a direct orientation. This orientation is a well defined tropism response." A similar statement had been made by

---

[d] Lillie also assumes that it is the intensity gradient which determines the direction of motion in tropistic reactions. This is not correct, since positively heliotropic animals go to the light even if by so doing they have to go from strong into weak light (see page 50). The direction of motion in tropistic reactions is determined by differences in the mass of chemical substances on both sides of a symmetrical animal.

Harper for the heliotropism of certain worms, namely that in strong light the animals move by heliotropism, in weak light by "trial and error." These statements are as erroneous as the assertion that while a stone falls under the influence of gravity a feather finds its way down by the method of "trial and error."

Barrows and Harper overlook the rôle of mass action and reaction velocity. When an animal is struck on one side only by light or by a chemically active substance emanating from a center of diffusion, the mass of this substance or of the photochemical reaction product increases on this side. These substances react with some substance of the nerve endings and as soon as the mass of the reaction product reaches a certain quantity the automatic turning, the tropistic reaction, occurs. When the light is strong or when the animal is near the center of diffusion, this happens in a short time and the tropistic character of the reaction is striking, since the animal is quickly put back into its proper orientation if it deviates from it. When the light is weak or when the animal is at some distance from the center of diffusion it will take a longer time before this critical value of the reaction product is reached, and in this case the animal can deviate considerably out of the correct orientation before it is brought back into the right orientation.

CHAPTER XVII

THERMOTROPISM

UNDER the name of thermotropism M. Mendelssohn [352-355] has described the observation that *Paramæcia* gather at a definite end of a trough when these ends have a different temperature. The organisms were put into a flat trough resting on tubes through which water was flowing. When the water in the tube had a temperature of 38° at one end of the trough, while the tube at the opposite end was perfused by water of 26° the organisms all gathered at the latter end. If then the temperature of the water in the two tubes was reversed the organisms went to the other end of the trough. If one end had the temperature of 10° the other of 25°, all went to the latter end. In this case we are in all probability not dealing with a tropistic reaction but with a collection of organisms due to the mechanism of motion described for *Paramæcium* by Jennings. When these organisms come suddenly from a region of a moderate temperature to one of lower temperature the activity of their cilia is transitorily reversed, but owing to the asymmetrical arrangement of their cilia they do not go back in the old direction but deviate to one side. This can lead to a collection of *Paramæcia* such as Mendelssohn described.

CHAPTER XVIII

INSTINCTS

THE teleological way of analyzing animal conduct has predominated to such an extent that there has been a tendency to connect all animal reactions with the preservation of the individual and the species. Instincts are considered to be such reactions of the organism as a whole which lead to the nutrition of the individual, the mating of the two sexes, and the care of the offspring. If the tropism theory of animal conduct is justified it must be possible to show that instincts are tropistic reactions.

We have insisted in previous chapters that animals indifferent to light can be made strongly positively or negatively heliotropic by certain chemicals or *vice versa* (*e.g.*, the experiments on certain fresh water crustaceans with acids or alcohol and caffein). We know that the body itself produces at various periods of its existence definite hormones and such hormones can act similarly as the acids or the caffein in the experiments on crustaceans, since it makes no difference whether such substances as acid are introduced into the blood from the outside or from certain tissues of the animal's own body. We know through F. Lillie's observations that in the blood of the male cattle embryo substances circulate which inhibit the development of secondary sexual characters of the female embryo, and we know through Steinach's experiments that the intermediate tissue from the sexual gland of one sex when introduced into the castrated organism of the opposite sex may impart to the latter the sexual instincts of the former. Hormones produced by definite tissues, therefore, influence the instincts. We want to show that this influence is due to a modification of *tropistic* reactions by the hormones.

Mating in certain fish, like *Fundulus,* consists in the male pressing that part of its body which contains the opening of the sperm duct against the corresponding part of the female body. The latter responds by pressing back, and the pressure of the body is maintained by both sexes through motions of the tail. During this mutual pressure or friction both sexes shed their sexual cells, sperm and eggs, into the water, and since the openings of the cloaca of the male and female, through which the sex cells are shed, are brought almost in contact with each other, sperm and eggs mix at the moment they are shed. This act of mating is due to a stereotropism which exists only during the spawning season and which is supposedly due to certain hormones existing at this time in the animal. The existence of such hormones is also indicated by certain colorations which develop and exist in the male during this period. This stereotropism is to some extent specific since it is exhibited by the contact between the two sexes. The specificity of this stereotropism is of importance and needs further experimental analysis, but that it is in reality a type of common stereotropism is evidenced by the fact that if during the spawning season we keep females isolated from males in an aquarium the females will go through the motions of mating and shed the eggs every time they come in contact with the glass walls of the aquarium. When they are kept permanently isolated from the male they repeat this non-specific purely stereotropic mating throughout the season. The eggs which they shed they quite frequently devour.

These manifestations of a highly developed stereotropism in the segments of the reproductive organs are probably widespread in the animal kingdom. The late Professor Whitman told the writer that male pigeons when kept in isolation will try to go through the motions of mating with any solid object in their field of vision, *e.g.*, glass bottles, and even with objects which give only the optical impression of a solid, namely, their own shadow on the ground.

In ants, the winged males and females become intensely positively heliotropic at the time of mating. Copulation occurs in the air, in the so-called nuptial flight. At a certain time—in the writer's observation toward sunset, when the sky is illuminated at the horizon only—the whole swarm of males and females leave the nest and fly in the direction of the glow. The wedding flight is a heliotropic phenomenon [287] presumably due to substances produced in the body during this period. After copulation the female loses its wings and also its positive heliotropism.[a] It becomes now intensely stereotropic. When kept in a dark box with pieces of cloth in folds the wingless female will now be found in the folds where its body is as closely as possible in contact with the solids. This positive stereotropism leads the queen to begin a subterranean existence which marks the founding of a new nest. Heliotropism and stereotropism are, therefore, the controlling factors in mating and the starting of a new nest in these ants.[287]

V. L. Kellogg [265] has made observations which show that the nuptial flight in bees is also due to an outburst of positive heliotropism as in the ant.

---

[a] It has already been mentioned that artificial removal of the wings of the fruit fly will also abolish its heliotropism.

In the course of some experiments on the sense-reactions of honey-bees, I have kept a small community of Italian bees in a glass-sided, narrow, high observation hive, so made that any particular bee, marked, which it is desired to observe constantly, can not escape this observation. The hive contains but two frames, one above the other, and is made wholly of glass, except for the wooden frame. It is kept covered, except during observation periods, by a black cloth jacket. The bees live contentedly and normally in this small hive, needing only occasional feeding at times when so many cells are given up for brood that there are not enough left for sufficient stored food supplies. Last spring at the normal swarming time, while standing near the jacketed hive, I heard the excited hum of a beginning swarm and noted the first issuers rushing pellmell from the entrance. Interested to see the behavior of the community in the hive during such an ecstatic condition as that of swarming, I lifted the cloth jacket, when the excited mass of bees which was pushing frantically down to the small exit in the lower corner of the hive turned with one accord about face and rushed directly upward away from the opening toward and to the top of the hive. Here the bees jammed, struggling violently. I slipped the jacket partly on; the ones covered turned down; the ones below stood undecided; I dropped the jacket completely; the mass began issuing from the exit again; I pulled off the jacket, and again the whole community of excited bees flowed—that is the word for it, so perfectly aligned and so evenly moving were all the individuals of the bee current—up to the closed top of the hive. Leaving the jacket off permanently, I prevented the issuing of the swarm until the ecstasy was passed and the usual quietly busy life of the hive was resumed. About three hours later there was a similar performance and failure to issue from the quickly unjacketed hive. On the next day another attempt to swarm was made, and after nearly an hour of struggling and moving up and down, depending on my manipulation of the black jacket, most of the bees got out of the hive's opening and the swarming came off on a weed bunch near the laboratory. That the issuance from the hive at swarming time depends upon a sudden extra-development of positive heliotropism seems obvious. The ecstasy comes and the bees crowd for the one spot of light in the normal hive, namely, the entrance opening. But when the covering jacket is lifted and the light comes strongly in from above—my hive was under a skylight—they rush toward the top, that is, toward the light. Jacket on and light shut off from above, down they rush; jacket off and light stronger from above than below and they respond like iron filings in front of an electromagnet which has its current suddenly turned on.

Finally there are indications of the rôle of chemotropism in mating. It has been observed for a long time that if a female butterfly is kept hidden from sight in a not too tightly closed box, male butterflies of the same species will be attracted by the box and settle on it. The female apparently gives off a substance to which the male is positively chemotropic. All these observations should be worked out more systematically. The data suffice, however, to indicate that what the biologist and psychologist call instinct are manifestations of tropisms.

The fact that eggs are laid by many insects on material which serves as a nutritive medium for the offspring is a typical instinct. An experimental analysis shows again that the underlying mechanism of the instinct is a positive chemotropism of the mother insect for the type of substance serving her as food; and when the intensity of these volatile substances is very high, *i.e.*, when the insect is on the material, the egg-laying mechanism of the fly is automatically set into motion. Thus the common housefly will deposit its eggs on decaying meat but not on fat; but it will also deposit it on objects smeared over with asafœtida, on which the larvæ cannot live. Aseptic banana flies will lay their eggs on sterile banana, although the banana is only an adequate food for the larvæ when yeast grows on it. It seems that the female insect lays her eggs on material for which she is positively chemotropic, and this is generally material which she also eats. The fact that such material serves as food for the coming generation is an accident. Considered in this way, the mystic aspect of the instinctive care of insects for the future generation is replaced by the simple mechanistic conception of a tropistic reaction. In this case natural selection plays a rôle since species whose females would too frequently lay their eggs on material on which the larvæ cannot thrive would be liable to die out.

As an illustration of the rôle of tropisms in the instinctive self-preservation the writer wishes to apologize for selecting an example which he has used so often in previous discussions, namely the rôle of heliotropism in the preservation of the life of the caterpillars of *Porthesia chrysorrhœa*.[287] This butterfly lays its eggs upon a shrub, on which the larvæ hatch in the fall and on which they hibernate, as a rule, not far from the ground. As soon as the temperature reaches a certain height, they leave the nest; under natural conditions this happens in the spring when the first leaves have begun to form on the shrub. (The larvæ can, however, be induced to leave the nest at any time in the winter, provided the temperature is raised sufficiently). After leaving the nest, they crawl directly upward on the shrub where they find the leaves on which they feed. If the caterpillars should move down the shrub they would starve, but this they never do, always crawling upward to where they find their food. What gives the caterpillar this never-failing certainty which saves its life and for which the human being might envy the little larva? Is it a dim recollection of experiences of former generations, as Samuel Butler would have us believe? It can be shown that this instinct is merely positive heliotropism and that the light reflected from the sky guides the animals upward. The caterpillars upon waking from their winter sleep are violently positively heliotropic, and it is this heliotropism which makes the animals move upward. At the top of the branch they come in contact with a growing bud and chemical and tactile influences set the mandibles of the young caterpillar into activity. If we put these caterpillars into closed test tubes which lie with their longitudinal axes at right angles to the window they will all migrate to the window end where they will stay and starve, even if we put their favorite leaves into the test tube close behind them. These larvæ are in this condition slaves of the light.

The few young leaves on top of a twig are quickly eaten by the caterpillar. The light which saved its life by making it creep upward where it finds its food would cause it to starve could the animal not free itself from the bondage of positive heliotropism. After having eaten, it is no longer a slave of light but can and does creep downward. It can be shown that a caterpillar after having been fed loses its positive heliotropism almost completely and permanently. If we submit unfed and fed caterpillars of the same nest to the same artificial or natural source of light in two different test tubes the unfed will creep to the light and stay there until they die, while those that have eaten will pay little or no attention to the light. Their positive heliotropism has disappeared and the animal after having eaten can creep in any direction. The restlessness which accompanies the condition of starvation makes the animal leave the top of the branches and creep downward—which is the only direction open to it—where it finds new young leaves on which it can feed. The wonderful hereditary instinct upon which the life of the animal depends is its positive heliotropism in the unfed condition and the loss of this heliotropism after having eaten. The chemical changes following the taking up of the food abolish the heliotropism just as CO_2 arouses positive heliotropism in certain *Daphnia*.

Mayer and Soule have shown that negative geotropism and positive heliotropism keep the caterpillars of *Danais plexippus* on its plant (the milk-weed). The chemical nature of the leaf starts the eating reactions, but "once the eating reaction be set into play, it tends to continue, so that the larva may then be induced to eat substances which it would never have commenced to eat in the first instance."[351]

These few examples may suffice to show that the theory of tropisms is at the same time the theory of instincts if due consideration is given to the rôle of hormones in producing certain tropisms and suppressing others. A systematic analysis of instinctive reactions from the viewpoint of the theory of tropisms and hormones will probably yield rich returns. As an example we may quote the fact that diurnal depth migrations of aquatic animals, consisting in an upward motion during the night and a downward motion during the day, are in all probability determined by a periodic change in the sense of heliotropism.[183, 300]

CHAPTER XIX

MEMORY IMAGES AND TROPISMS

When a muscle is stimulated several times in succession, the effect of the second or third or later stimulation may be greater than that of the first. A consistently anthropomorphic author should draw the inference that the muscle is gradually learning to react properly. What seems to happen is that the hydrogen ion concentration is raised by the first stimulations to a point where the effect of the stimulation becomes greater. When the stimulations continue and the hydrogen ion concentration becomes still greater, the response of the muscle declines and finally becomes zero; the hydrogen ion concentration has now become too high. The writer observed that when winged plant lice of a *Cineraria* were taken directly from the plant, they did not react as promptly as after they had gone through several heliotropic experiments. There is nothing to indicate that this is a case of "learning," since it may also be the result of a change in the hydrogen ion concentration or of some other reaction product. It may also be the result of some purely mechanical obstacle to rapid locomotion being removed.

We can speak of learning only in such organisms in which the existence of associative memory can be proved. By associative memory we mean that mechanism, by which a stimulus produces not only the direct effects determined by its nature, but also the effects of entirely different stimuli which at some former period by chance attacked the organism at the same time with the given

MEMORY IMAGES

stimulus. Thus the image or the odor of a rose may call up the memory of persons or surroundings which were present on a former occasion when the image or odor of the flower impressed us. Brain physiology shows that this type of associative memory is the specific function of definite parts of the brain, e.g., the cerebral hemispheres which exist only in definite types of animals. We see also that certain species among vertebrates, insects, crustacea, and cephalopods possess associative memory, while to the knowledge of the writer no adequate proof for its existence has ever been given for worms, starfish, sea urchins, actinians, medusæ, hydroids, or infusorians.[293] Claims for the existence of such memory in these latter groups of animals have frequently been made, but such claims are either plain romance or due to a confusion of reversible physiological processes with the irreversible phenomena of associative memory. The less a scientist is accustomed to rigid quantitative experiments, the more ready he is to confound the reversible after effects of a stimulus—e.g., the after effects due to an increase in hydrogen ion concentration—with indications of associative memory. Learning is only possible where there exists a specific organ of associative memory, the physical mechanism of which is still unknown.

The manifestations of associative memory are generally discussed by the introspective psychologists, who as a rule are not familiar with or do not appreciate the methods of the physicist. There have been made repeated attempts to develop methods for the analysis of associative memory, among which thus far only one satisfies the demands of quantitative science, namely Pawlow's method. As is well known even to the layman, eating causes a flow of saliva. The quantity of saliva excreted by the parotid (one of the salivary glands) in the dog can be collected and measured. The earlier physiological workers had observed that in a dog which had often been used for the study of the influence of eating upon the flow of saliva, the saliva began to flow whenever the preparations for feeding were made before the eyes of the dog, even when no food was given. Pawlow made use of this fact to study quantitatively the "strength" of such associative phenomena, which he terms "conditioned reflexes" (to escape the terminology and interpretations of the introspective psychologist).[537] A fistula [a] of the duct of the parotic gland allows the saliva to flow outside the cavity of the mouth. This fistula is connected with a long manometer which by a special air chamber arrangement gives a considerable change in the height of the meniscus for the secretion of as little as one drop of saliva. The variations of the height of the column of liquid in the manometer are observed outside of the room where the dog is. For each dog which is to serve for such experiments the meal is preceded by a certain signal, the sounds of a metronome of definite rhythm, or a definite musical sound, or a definite optical signal, and so forth, which is to form the special conditioned reflex for this dog. After a certain number of repetitions the association is established and from now on the flow of saliva commences from the dog's parotid when the typical signal is given. It was found that the quantity of saliva excreted by the signal changes in a definite sense and quantity when the signal varies or when other conditions accompanying the signal vary.

---

[a] The writer is indebted for the details of Pawlow's method to a short review by Dr. Morgulis.[537, 538]

Thus in one dog "by persistent training a conditioned reflex has been established to the stimulation with 100 oscillations per minute of the metronome. The stimulation of intermittent sounds of such frequency called forth 6 to 10 drops of saliva every time. The interval between successive oscillations was then modified, the moment of the disappearance of the conditioned salivary reflex indicating the lowest limit of differentiation. Without going into any details of this most interesting investigation or quoting actual data, I will say that the dog could sharply distinguish the shortening of the interval by less than 1/40 to 1/43 of a second. Indeed with the well-developed reflex to the stimulation of 100 beats per minute a change of the rate to either 96 or 104 beats was immediately reacted upon by a marked diminution or even complete cessation of the flow of saliva."

This example will give an indication how sensitive is this method of measuring the effect of a memory association.

It is not our purpose to give the details of Pawlow's results—they have only been published in Russian and are therefore not accessible to the writer—but to show that the influence of an associative memory image is as exactly measurable as, *e.g.,* the direct illumination of the eye; and moreover that what we call a memory image is not a "spiritual" but a physical agency. We therefore need not be surprised to find that such memory images or "conditioned reflexes" can vary and multiply the number of possible tropistic reactions.

We have mentioned in the previous chapter that the stereotropism in the mating instinct includes apparently an element of species specificity inasmuch as naturally only males and females of the same species mate. The late Professor Whitman has shown by experiment that this specificity is, in pigeons at least, not inherited but the effect of memory images (a "conditioned reflex" in the sense of Pawlow). Whitman took the eggs or young of wild species, giving them to the domestic ring-dove to foster, with the result, that the young reared by the ring-doves ever after associated with ring-doves and tried to mate with them. Passenger pigeons when reared by ring-doves refuse to mate with their own species but mate with the species of the foster parents.[539] This shows incidentally that racial antagonism is not inherited but acquired.

We have mentioned the fact that the mating instinct is determined by tropisms aroused by specific internal secretions, and that in isolated male pigeons any solid body can arouse the mating reaction. Craig[540] raised male pigeons in isolation so that they never came in contact with other pigeons until they were adult. One pigeon was hatched in July and isolated in August.

> Throughout the autumn and early winter this bird cooed very little. But about the first of February there began a remarkable development of voice and social behavior. The dove was kept in a room where several men were at work, and he directed his display behavior toward these men just as if they belonged to his own species. Each time I put food in his cage he became greatly excited, charging up and down the cage, bowing-and-cooing to me, and pecking my hand whenever it came within his cage. From that day until the day of his death, Jack continued to react in this social manner to human beings. He would bow-and-coo to me at a distance, or to my face when near the cage; but he paid greatest attention to the hand—naturally so, because it was the only part with which he daily came into direct contact. He treated the hand much as if it were a living bird. Not only were his own activities directed toward the hand as if it were a bird, but he received treatment by the hand in the same spirit. The hand could stroke him, preen his neck, even pull the feathers sharply, Jack had absolutely no fear, but ran to the hand to be stroked or teased, showing the joy that all doves show in the attentions of their companions.

When this pigeon was almost a year old it was put into a cage with a female pigeon, but although the female aroused the sexual instinct of the formerly isolated male the latter did not mate with her, but mated with the hand of his attendant when the hand was put into the cage, and this continued throughout the season. Thus the memory images acquired by the bird at an impressionable age and period perverted its sexual tropisms.

It is perhaps of more importance to show that memory images may have a direct orienting influence. The chemotropic phenomenon of an insect laying its egg on a substance which serves as food (for both mother and offspring) and for which the mother is positively chemotropic, may be modified by an act of associative memory, *e.g.*, when a solitary wasp drags the caterpillar on which it lays its eggs to a previously prepared hole in the ground. The essential part of the instinct, the laying of the eggs on the caterpillar, does, perhaps, not differ very much from the fly laying its eggs on decaying meat; and the solitary wasp may be strongly positively chemotropic for the caterpillar on which it lays the eggs, although this has not yet been investigated. But the phenomenon is complicated by a second tropism, which we will call the orienting effect of the memory image. As is well known, the wasp before "going for" the caterpillar digs a hole in the ground to which it afterwards drags the caterpillar, often from a distance. The finding of this previously prepared hole by the returning wasp, the writer would designate as the tropistic or orienting effect of the memory image of the location of this hole; meaning thereby that the memory image of the location of this hole makes the animal return to this location. The conduct of these wasps is familiar to many readers and the writer may be pardoned for quoting from a formerly published observation.

> *Ammophila,* a solitary wasp, makes a small hole in the ground and then goes out to hunt for a caterpillar, which, when found, it paralyses by one or several stings. The wasp carries the caterpillar back to the nest, puts it into the hole, and covers the latter with sand. Before this is done, it deposits its eggs on the caterpillar which serves the young larva as food.
>
> An *Ammophila* had made a hole in a flower bed and left the flower bed flying. A little later I saw an *Ammophila* running on the sidewalk of the street in front of the garden, dragging a caterpillar which it held in its mouth. The weight of the caterpillar prevented the wasp from flying. The garden was higher than the sidewalk and separated from it by a stone wall. The wasp repeatedly made an attempt to climb upon the stone wall, but kept falling down. Suspecting that it might have a hole prepared in the garden, I was curious to see whether and how it would find the hole. It followed the wall until it reached the neighboring yard, which had no wall. It now left the street and crawled into this yard, dragging the caterpillar along. Then crawling through the fence which separated the two yards, it dropped the caterpillar near the foot of a tree, and flew away. After a short zigzag flight it alighted on a flower bed in which I noticed two small holes. It soon left the bed and flew back to the tree, not in a straight line but in three stages, stopping twice on its way. At the third stop it landed at the place where the caterpillar lay. The caterpillar was then dragged to the hole, pulled into it, and the hole was covered with tiny stones in the usual way.[293]

It is not enough to say that the animal possesses associative memory and returns to the hole; we must add that the brain image of the region of the hole becomes the source of a forced orientation of the animal—of an added special tropism—compelling the animal to return to the region corresponding to the image. And the same may be said in regard to the return of the wasp to the caterpillar which had been temporarily deposited at the foot of the tree.

This example, which might be easily multiplied, will

show the addition necessary to the tropism theory to make it include the endless number of reactions in which associative memory is involved. The psychiatrist would find it easy to supply numerous examples of this type of forced movements toward certain objects which have left a memory image. Since the writer has not investigated this subject sufficiently he is not in a position to give more than a suggestion for the direction of further work. He is inclined to believe that with this enlargement the tropism theory might include human conduct also if we realize that certain memory images may exercise as definite an orienting influence as, *e.g.*, moving retina images or sex hormones.

This tentative extension of the forced movement or tropism theory of animal conduct may explain why higher animals and human beings seem to possess freedom of will, although all movements are of the nature of forced movements. The tropistic effects of memory images and the modification and inhibition of tropisms by memory images make the number of possible reactions so great that prediction becomes almost impossible and it is this impossibility chiefly which gives rise to the doctrine of free will. The theory of free will originated and is held not among physicists but among verbalists. We have shown that an organism goes where its legs carry it and that the direction of the motion is forced upon the organism. When the orienting force is obvious to us, the motion appears as being willed or instinctive; the latter generally when all individuals act alike, machine fashion, the former when different individuals act differently. When a swarm of *Daphnia* is sensitized with CO_2 they all rush to the source of light. This is a machine-like action, and many will be willing to admit that it is a forced movement or an instinctive reaction. After the CO_2 has evaporated the animals become indifferent to light, and while formerly they had only one degree of freedom of motion they now can move in any direction. In this case the motions appear to be spontaneous or free, since we are not in a position to state why *Daphnia a* moves to the right and *Daphnia b* to the left, etc. As a matter of fact, the motion of each individual is again determined by something but we do not know what it is. The persistent courtship of a human male for a definite individual female may appear as an example of persistent will, yet it is a complicated tropism in which sex hormones and definite memory images are the determining factors. Removal of the sex glands abolishes the courtship and replacing the sex glands of an individual by those of the opposite sex may lead to a complete reversal of the sex instincts. What appears as persistent will action is, therefore, essentially a tropistic reaction. The production of heliotropism by CO_2 in *Daphnia* and the production of the definite courtship of the male A for the female B are similar phenomena differing only by the nature of the hormones and the additional tropistic effects of certain memory images in the case of courtship. Our conception of the existence of "free will" in human beings rests on the fact that our knowledge is often not sufficiently complete to account for the orienting forces, especially when we carry out a "premeditated" act, or when we carry out an act which gives us pain or may lead to our destruction, and our incomplete knowledge is due to the sheer endless number of possible combinations and mutual inhibitions of the orienting effect of individual memory images.

LITERATURE [a]

[1] ABBOTT, J. F., and LIFE, A. C.: Galvanotropism in Bacteria. *Am. J. Physiol.*, 1908, xxii, 202–206.
[2] ADAMS, G. P.: On the Negative and Positive Phototropism of the Earthworm *Allolobophora fœtida* (Sav.) as Determined by Light of Different Intensities. *Am. J. Physiol.*, 1903, ix, 26–34.
[3] ALLEE, W. C.: An Experimental Analysis of the Relation Between Physiological States and Rheotaxis in Isopoda. *J. Exp. Zool.*, 1912, xiii, 269–344.
[4] ALLEE, W. C.: The Effect of Molting on Rheotaxis in Isopods. *Science*, 1913, xxxvii, 882–883.
[5] ALLEE, W. C.: Further Studies on Physiological States and Rheotaxis in Isopoda. *J. Exp. Zool.*, 1913, xv, 257–295.
[6] ALLEE, W. C.: Certain Relations Between Rheotaxis and Resistance to Potassium Cyanide in Isopoda. *J. Exp. Zool.*, 1914, xvi, 397–412.
[7] ALLEE, W. C.: The Ecological Importance of the Rheotactic Reaction of Stream Isopods. *Biol. Bull.*, 1914, xxvii, 52–66.
[8] ALLEE, W. C.: Chemical Control of Rheotaxis in *Asellus*. *J. Exp. Zool.*, 1916, xxi, 163–198.
[9] ALLEE, W. C., and TASHIRO, S.: Some Relations Between Rheotaxis and the Rate of Carbon Dioxide Production of Isopods. *J. Animal Behav.*, 1914, iv, 202–214.
[10] ALLEN, G. D.: Reversibility of the Reactions of *Planaria dorotocephala* to a Current of Water. *Biol. Bull.*, 1915, xxix, 111–128.
[11] AREY, L. B.: The Orientation of *Amphioxus* During Locomotion. *J. Exp. Zool.*, 1915, xix, 37–44.
[12] ARISZ, W. H.: On the Connection Between Stimulus and Effect in Phototropic Curvatures of Seedlings of *Avena sativa*. *Proc. Roy. Acad. Amsterdam*, 1911, xiii, 1022–1031.
[13] AXENFELD, D.: Quelques observations sur la vue des arthropodes. *Arch. Ital. Biol.*, 1899, xxxi.
[14] BACH, H.: Ueber die Abhängigkeit der geotropischen Präsentations- und Reaktionszeit von verschiedenen äusseren Faktoren. *Jahrb. wiss. Bot.*, 1907, xliv, 57–123.
[15] BALSS, H.: Uber die Chemorezeption bei Garneelen. *Biol. Centr.*, 1913, xxxiii, 508–512.

---

[a] The list of literature does not claim to be complete. Aside from unintentional omissions, some of the controversial and amateurish publications have not been included.

[16] Bancroft, F. W.: Note on the Galvanotropic Reactions of the Medusa *Polyorchis penicillata*, A. Agassiz. *J. Exp. Zool.*, 1904, i, 289–292.

[17] Bancroft, F. W.: Ueber die Gültigkeit des Pflüger'schen Gesetzes für die galvanotropischen Reaktionen von *Paramæcium*. *Arch. ges. Physiol.*, 1905, cvii, 535–556.

[18] Bancroft, F. W.: On the Influence of the Relative Concentration of Calcium Ions on the Reversal of the Polar Effects of the Galvanic Current in *Paramæcium*. *J. Physiol.*, 1906, xxxiv, 444–463.

[19] Bancroft, F. W.: The Control of Galvanotropism in *Paramæcium* by Chemical Substances. *Univ. Cal. Pub. Physiol.*, 1906, iii, 21–31.

[20] Bancroft, F. W.: The Mechanism of Galvanotropic Orientation in *Volvox*. *J. Exp. Zool.*, 1907, iv, 157–163.

[21] Bancroft, F. W.: Heliotropism, Differential Sensibility, and Galvanotropism in *Euglena*. *J. Exp. Zool.*, 1913, xv, 383–428.

[22] Banta, A. M.: Experiments on the Light and Tactile Reactions of a Cave Variety and an Open Water Variety of an Amphipod Species. *Proc. Soc. Exp. Biol. and Med.*, 1913, x, 192.

[23] Baranetzki, J.: Influence de la lumière sur les plasmodia des myxomycètes. *Mém. Soc. Sc. Nat. Cherbourg*, 1876, xix, 321–360.

[24] Barratt, J. O. W.: Der Einfluss der Konzentration auf die Chemotaxis. *Z. allg. Physiol.*, 1905, v, 73–94.

[25] Barrows, W. M.: The Reactions of the Pomace Fly, *Drosophila ampelophila* Loew, to Odorous Substances. *J. Exp. Zool.*, 1907, iv, 515–537.

[26] Bauer, V.: Ueber die reflektorische Regulierung der Schwimmbewegungen bei den Mysiden, mit besonderer Berücksichtigung der doppelsinnigen Reizbarkeit der Augen. *Z. allg. Physiol.*, 1908, viii, 343–370.

[27] Bauer, V.: Ueber sukzessiven Helligkeitskontrast bei Fischen. *Centr. Physiol.*, 1909, xxiii, 593–599.

[28] Bauer, V.: Ueber das Farbenunterscheidungsvermögen der Fische. *Arch. ges. Physiol.*, 1910, cxxxiii, 7–26.

[29] Bauer, V.: Zur Kenntnis der Lebensweise von *Pecten jacobæus* L. Im besonderen über die Funktion der Augen. *Zool. Jahrb. Abt. allg. Zool.*, 1912, xxxiii, 127–150.

[30] Baunacke, W.: Statische Sinnesorgane bei den Nepiden. *Zool. Jahrb. Abt. Anat.*, 1912–13, xxxiv, 179–346.

[31] Baunacke, W.: Studien zur Frage nach der Statocystenfunktion. I. Statische Reflexe bei Mollusken. *Biol. Centr.*, 1913, xxxiii, 427–452.

LITERATURE

32 BAUNACKE, W.: II. Noch einmal die Geotaxis unserer Mollusken. *Biol. Centr.*, 1914, xxxiv, 371–385; 497–523.

33 BEER, TH.: Vergleichend-physiologische Studien zur Statocystenfunktion. I. Ueber den angeblichen Gehörsinn und das angebliche Gehörorgan der Crustaceen. *Arch. ges. Physiol.*, 1898, lxxiii, 1–41.

34 BEER, TH.: II. Versuche an Crustaceen (*Penæus membranaceus*). *Arch. ges. Physiol.*, 1899, lxxiv, 364–382.

35 BENGT, J.: Der richtende Einfluss strömenden Wassers auf wachsende Pflanzen und Pflanzenteile (Rheotropismus). *Ber. bot. Ges.*, 1883, i, 512–521.

36 BERNSTEIN, J.: Chemotropische Bewegung eines Quecksilbertropfens. Zur Theorie der amöboiden Bewegung. *Arch. ges. Physiol.*, 1900, lxxx, 628–637.

37 BERT, P.: Les animaux voient-ils les mêmes rayons lumineux que nous? *Mém. Soc. Sc. Phys. et Nat. Bordeaux*, 1868, vi, 375–383.

38 BERT, P.: Sur la question de savoir si tous les animaux voient les mêmes rayons lumineux que nous. *Arch. de Physiol.*, 1869, ii, 547–554.

39 BETHE, A.: Ueber die Erhaltung des Gleichgewichts. *Biol. Centr.*, 1894, xiv, 95–114; 563–582.

40 BETHE, A.:Die Otocyste von Mysis. *Zool. Jahrb. Abt. Anat.*, 1895, viii, 544–564.

41 BETHE, A.: Die Locomotion des Haifisches (*Scyllium*) und ihre Beziehungen zu den einzelnen Gehirnteilen und zum Labyrinth. *Arch. ges. Physiol.*, 1899, lxxvi, 470–493.

42 BIRGE, E. A.: The Vertical Distribution of the Limnetic Crustacea of Lake Mendota. *Biol. Centr.*, 1897, xvii, 371–374.

43 BIRUKOFF, B.: Untersuchungen über Galvanotaxis. *Arch. ges. Physiol.*, 1899, lxxvii, 555–585.

44 BIRUKOFF, B.: Zur Theorie der Galvanotaxis. *Arch. Anat. u. Physiol., Physiol. Abt.*, 1904, 271–296.

45 BIRUKOFF, B.: Zur Theorie der Galvanotaxis. II. *Arch. ges. Physiol.*, 1906, cxi, 95–143.

46 BLAAUW, A. H.: The Intensity of Light and the Length of Illumination in the Phototropic Curvature in Seedlings of *Avena sativa* (Oats). *Proc. Roy. Akad. Amsterdam*, 1908.

47 BLAAUW, A. H.: Die Perzeption des Lichtes. *Rec. trav. bot. Neérlandais*, 1909, v, 209–377.

48 BLAAUW, A. H.: Licht und Wachstum. *Z. Bot.*, 1914, vi, 641–703; 1915, vii, 465–532.

48a BLASIUS, E., and SCHWEIZER, F.: Elektrotropismus und verwandte Erscheinungen. *Arch. ges. Physiol.*, 1893, liii, 493–543.

[49] Bohn, G.: Les *Convoluta roscoffensis* et la théorie des causes actuelles. *Bull. Mus. Paris,* 1903, 352–364.

[50] Bohn, G.: Théorie nouvelle du phototropisme. *Compt. rend. Acad. Sc.,* 1904, cxxxix, 890–891.

[51] Bohn, G.: Attractions et oscillations des animaux marins sous l'influence de la lumière. Recherches nouvelles relatives au phototactisme et au phototropisme. *Mém. Inst. génér. Psychol.,* 1905, i, 1–111.

[52] Bohn, G.: Impulsions motrices d'origine oculaire chez les crustacés. (Deuxieme mémoire relatif au phototactisme et au phototropisme.) *Bull. Inst. génér. Psychol.,* 1905, v, 412–454.

[53] Bohn, G.: Intervention des réactions oscillatoires dans les tropismes. *Compt. rend. Assoc. Française avancement des Sc., Congrès de Reims,* 1907, 700–706.

[54] Bohn, G.: Observations biologiques sur le branchellion de la torpille. *Bull. Station biol. Arcachon,* 1907, x, 283–296.

[55] Bohn, G.: Les tropismes, la sensibilité différentielle et les associations chez le branchellion de la torpille. *Compt. rend. Soc. Biol.,* 1907, lxiii, 545–548.

[56] Bohn, G.: A propos des lois de l'excitabilité par la lumière. I. Le retour progressif à l'état d'immobilité, après une stimulation mécanique. *Compt. rend. Soc. Biol.,* 1907, lxiii, 655–658.

[57] Bohn, G.: II. Du changement de signe du phototropisme en tant que manifestation de la sensibilité différentielle. *Compt. rend. Soc. Biol.,* 1907, lxiii, 756–759.

[58] Bohn, G.: Introduction à la psychologie des animaux a symétrie rayonnée. I. Les états physiologiques des actinies. *Bull. Inst. génér. Psychol.,* 1907, vii, 81–129; 135–182.

[59] Bohn, G.: II. Les essais et erreurs chez les étoiles de mer et les ophiures. *Bull. Inst. génér. Psychol.,* 1908, viii, 21–102.

[60] Bohn, G.: Les rythmes vitaux chez les actinies. *Compt. rend. Assoc. Française avancement des Sc.,* 1908, 613.

[61] Bohn, G.: De l'orientation chez les patelles. *Compt. rend. Acad. Sc.,* 1909, cxlviii, 868–870.

[62] Bohn, G.: Les variations de la sensibilité périphérique chez les animaux. *Bull. Sc. France et Belgique,* 1909, xliii, 481–519.

[63] Bohn, G.: Quelques problèmes généraux relatifs à l'activité des animaux inférieurs. *Bull. Inst. génér. Psychol.,* 1909, ix, 439–466.

[64] Bohn, G.: Quelques observations sur les chenilles des dunes. *Bull. Inst. génér. Psychol.,* 1909, ix, 543–549.

[65] Bohn, G.: La naissance de l'intelligence. Paris, 1909.

66 BOHN, G.: Les tropismes. *Rapport VIme Congr. Internat. Psychol. Genève,* 1909, pp. 15.
67 BOHN, G.: A propos les lois de l'excitabilité par la lumière. III. De l'influence de l'éclairement du fond sur le signe des réactions vis-à-vis la lumière. *Compt. rend. Soc. Biol.,* 1909, lxvi, 18–20.
68 BOHN, G.: IV. Sur les changements périodiques du signe des réactions. *Compt. rend. Soc. Biol.,* 1909, lxvii, 4–6.
69 BOHN, G.: V. Intervention de la vitesse des réactions chimiques dans la désensibilisation par la lumière. *Compt. rend. Soc. Biol.,* 1910, lxviii, 1114–1117.
70 BOHN, G.: La sensibilisation et la désensibilisation des animaux. *Compt. rend. Assoc. Française avancement des Sc., Congrès de Toulouse,* 1910, 214–222.
71 BOHN, G.: Quelques expériences de modification des réactions chez les animaux, suivies de considérations sur les mécanismes chimiques de l'évolution. *Bull. Sc. France et Belgique,* 1911, xlv, 217–238.
72 BOHN, G.: La nouvelle psychologie animale. Paris, 1911, pp. 200.
73 BOHN, G.: La sensibilité des animaux aux variations de pression. *Compt. rend. Acad. Sc.,* 1912, cliv, 240–242.
74 BOHN, G.: Les variations de la sensibilité en relation avec les variations de l'état chimique interne. *Compt. rend. Acad. Sc.,* 1912, cliv, 388–391.
75 BOHN, G.: L'étude des phénomènes mnémiques chez les organismes inférieurs. *J. Psychol. u. Neurol.,* 1913, xx, 199–209.
76 BORING, E. G.: Note on the Negative Reaction Under Light Adaptation in the *Planarian. J. Animal Behav.,* 1912, ii, 229–248.
77 BORN, G.: Biologische Untersuchungen. Ueber den Einfluss der Schwere auf das Froschei. *Arch. mikr. Anat.,* 1885, xxiv, 475.
78 BREUER, J.: Ueber die Funktion der Bogengänge des Ohrlabyrinths. *Med. Jahrb.,* 1874.
79 BREUER, J.: Beiträge zur Lehre vom statischen Sinne. *Med. Jahrb.,* 1875.
80 BREUER, J.: Ueber die Funktion der Otolithenapparate. *Arch. ges. Physiol.,* 1891, xlviii, 195–306.
80a BREUER, J.: Ueber den Galvanotropismus (Galvanotaxis) bei Fischen. *Sitzngsb. Akad. Wiss. Wien. mathem.-naturw. Kl.,* 1905, cxiv, 27–56.
80b BREUER, J., and KREIDL, A.: Ueber die scheinbare Drehung des Gesichtsfeldes, während der Einwirkung einer Centrifugalkraft. *Arch. ges. Physiol.,* 1898, lxx, 494–510.
81 BRUCHMANN, H.: Chemotaxis der *Lycopodium*-Spermatozoiden. *Flora,* 1908–09, xcix, 193–202.

[82] BRUN, R.: Die Raumorientierung der Ameisen und das Orientierungsproblem im allgemeinen. Jena, 1914, pp. 242.

[83] BRUNDIN, T. M.: Light Reactions of Terrestrial Amphipods. *J. Animal Behav.*, 1913, iii, 334–352.

[84] v. BUDDENBROCK, W.: Untersuchungen über die Schwimmbewegungen und die Statocysten der Gattung *Pecten*. *Sitzngsb. Heidelberger Akad. Wiss., mathem.-naturw. Kl.*, 1911, pp. 24.

[85] v. BUDDENBROCK, W.: Ueber die Funktion der Statocysten im Sande grabender Meerestiere (*Arenicola* und *Synapta*). *Biol. Centr.*, 1912, xxxii, 564–585.

[86] v. BUDDENBROCK, W.: Ueber die Funktion der Statocysten von *Branchiomma vesiculosum*. *Verhandl. naturhist.-med. Vereines, Heidelberg*, 1913, N.F. xii, 256–261.

[87] v. BUDDENBROCK, W.: Ueber die Orientierung der Krebse im Raum. *Zool. Jahrb. Abt. Zool.*, 1914, xxxiv, 479–514.

[88] v. BUDDENBROCK, W.: A Criticism of the Tropism Theory of Jacques Loeb. *J. Animal Behav.*, 1916, vi, 341–366.

[89] BULLER, A. H. R.: Contributions to Our Knowledge of the Physiology of the Spermatozoa of Ferns. *Annals Bot.*, 1900, xiv, 543–582.

[90] BULLER, A. H. R.: Is Chemotaxis a Factor in the Fertilization of the Eggs of Animals? *Quart. J. Micr. Sc.*, 1902–03, xlvi, 145–176.

[91] BOYSEN-JENSEN, P.: Ueber die Leitung des phototropischen Reizes in der *Avena*koleoptile. *Ber. bot. Ges.*, 1913, xxxi, 559–566.

[92] BUNTING, M.: Ueber die Bedeutung der Otolithenorgane für die geotropischen Funktionen von *Astacus fluviatilis*. *Arch. ges. Physiol.*, 1893, liv, 531–537.

[93] CARLGREN, O.: Der Galvanotropismus und die innere Kataphorese. *Z. allg. Physiol.*, 1905, v, 123–130.

[94] CARLGREN, O.: Ueber die Einwirkung des konstanten galvanischen Stromes auf niedere Organismen. *Arch. Anat. u. Physiol., Physiol. Abt.*, 1900, 49–76.

[95] CARPENTER, F. W.: The Reactions of the Pomace Fly (*Drosophila ampelophila*, Loew) to Light, Gravity, and Mechanical Stimulation. *Am. Nat.*, 1905, xxxix, 157–171.

[95a] CLAPARÈDE, E.: Les tropismes devant la psychologie. *J. Psychol. u. Neurol.*, 1908, xiii, 150–160.

[96] CLARK, G. P.: On the Relation of the Otocysts to Equilibrium Phenomena in *Gelasimus pugilator* and *Platyonichus ocellatus*. *J. Physiol.*, 1896, xix, 327–343.

[97] CLARK, O. L.: Ueber negativen Phototropismus bei *Avena sativa*. *Z. Bot.*, 1913, v, 737–770.

[98] COEHN, A., and BARRATT, W.: Ueber Galvanotaxis vom Standpunkte der physikalischen Chemie. *Z. allg. Physiol.*, 1905, v, 1–9.

[99] COHN, F.: Ueber die Gesetze der Bewegung mikroskopischer Tiere und Pflanzen unter Einfluss des Lichtes. *Jahr.-ber. Schles. Ges. vaterl. Kultur*, 1864, xlii, 35–36.

[100] COLE, L. J.: The Influence of Direction *vs.* Intensity of Light in Determining the Phototropic Responses of Organisms. *Science*, 1907, xxv, 784.

[101] CONGDON, E. D.: Recent Studies Upon the Locomotor Responses of Animals to White Light. *J. Comp. Neurol. and Psychol.*, 1908, xviii, 309–328.

[102] CORNETZ, V.: Ueber den Gebrauch des Ausdruckes "tropisch" und über den Charakter der Richtungskraft bei Ameisen. *Arch. ges. Physiol.*, 1912, cxlvii, 215–233.

[103] COWLES, R. P.: Stimuli Produced by Light and by Contact with Solid Walls as Factors in the Behavior of Ophiuroids. *J. Exp. Zool.*, 1910, ix, 387–416.

[104] COWLES, R. P.: Reaction to Light and Other Points in the Behavior of the Starfish. *Papers from Tortugas Lab. Carnegie Inst. Washington*, 1911, iii, 95–110.

[105] COWLES, R. P.: The Influence of White and Black Walls on the Direction of Locomotion of the Starfish. *J. Animal Behav.*, 1914, iv, 380–382.

[105a] CRAIG, W.: The Voices of Pigeons Regarded as a Means of Social Control. *Am. J. Sociology*, 1908, xiv, 86–100.

[105b] CRAIG, W.: Male Doves Reared in Isolation. *J. Animal Behav.*, 1914, iv, 121–133.

[105c] CRAIG, W.: Appetites and Aversions as Constituents of Instincts. *Biol. Bull.*, 1918, xxxiv, 97–107.

[106] CROZIER, W. J.: The Orientation of a Holothurian by Light. *Am. J. Physiol.*, 1914, xxxvi, 8–20.

[107] CROZIER, W. J.: The Behavior of Holothurians in Balanced Illumination. *Am. J. Physiol.*, 1917, xliii, 510–513.

[108] CROZIER, W. J.: The Photoreceptors of *Amphioxus*. *Anat. Rec.*, 1917, xi, 520.

[108a] CROZIER, W. J.: The Photic Sensitivity of *Balanoglossus*. *J. Exp. Zool.*, 1917, xxiv, 211–217.

[109] CZAPEK, F.: Ueber Zusammenwirken von Heliotropismus und Geotropismus. *Sitzngsb. Akad. Wiss. Wien. mathem.-naturw. Kl.*, 1895, civ.

[110] CZAPEK, F.: Untersuchungen über Geotropismus. *Jahrb. wiss. Bot.*, 1895, xxvii, 243–339.

[111] CZAPEK, F.: Weitere Beiträge zur Kenntnis der geotropischen Reizbewegungen. *Jahrb. wiss. Bot.*, 1898, xxxii, 175–308.
[112] DALE, H. H.: Galvanotaxis and Chemotaxis of Ciliate Infusoria. *J. Physiol.*, 1901, xxvi, 291–361.
[113] DAVENPORT, C. B.: Experimental Morphology. Part I. Effects of Chemical and Physical Agents Upon Protoplasm. New York, 1897.
[114] DAVENPORT, C. B., and CANNON, W. B.: On the Determination of the Direction and Rate of Movement of Organisms by Light. *J. Physiol.*, 1897, xxi, 22–32.
[115] DAVENPORT, C. B., and LEWIS, F. T.: Phototaxis of *Daphnia*. *Science*, 1899, ix, 368.
[116] DAVENPORT, C. B., and PERKINS, H.: A Contribution to the Study of Geotaxis in the Higher Animals. *J. Physiol.*, 1897, xxii, 99–110.
[117] DAY, E. C.: The Effect of Colored Light on Pigment Migration in the Eye of the Crayfish. *Bull. Mus. Comp. Zool.*, 1911, liii, 303–343.
[118] DELAGE, Y.: Étude expérimentale sur les illusions statiques et dynamiques de direction pour servir à determiner les fonctions des canaux semicirculaires de l'oreille interne. *Arch. Zool. expér. et génér.*, 1886, (2) iv.
[119] DELAGE, Y.: Sur une fonction nouvelle des otocystes comme organes d'orientation locomotrice. *Arch. Zool. expér. et génér.*, 1887, (2) v, 1–26.
[120] DEWITZ, J.: Ueber die Vereinigung der Spermatozoen mit dem Ei. *Arch. ges. Physiol.*, 1885, xxxvii, 219–223.
[121] DEWITZ, J.: Ueber Gesetzmässigkeit in der Ortsveränderung der Spermatozoen und in der Vereinigung derselben mit dem Ei. *Arch. ges. Physiol.*, 1886, xxxviii, 358–385.
[122] DEWITZ, J.: Ueber den Rheotropismus bei Tieren. *Arch. Physiol.*, 1899 (Suppl.), 231–244.
[123] DOLLEY, W. L., JR.: Reactions to Light in *Vanessa antiopa*, with Special Reference to Circus Movements. *J. Exp. Zool.*, 1916, xx, 357–420.
[124] DRIESCH, H.: Heliotropismus bei Hydroidpolypen. *Zool. Jahrb.*, 1890, v, 147–156.
[125] DRIESCH, H.: Die taktische Reizbarkeit der Mesenchymzellen von *Echinus microtuberculatus*. *Arch. Entwcklngsmech.*, 1896, iii, 362–380.
[126] DRIESCH, H.: Die organischen Regulationen. Leipzig, 1901, pp. 228.
[127] DUBOIS, R.: Sur le mécanisme des fonctions photodermatique et photogénique dans le siphon du *Pholas dactylus*. *Compt. rend. Acad. Sc.*, 1889, cix, 233–235.

128 DUBOIS, R.: Sur l'action des agents modificateurs de la contraction photodermatique chez le *Pholas dactylus*. *Compt. rend. Acad. Sc.*, 1898, cix, 320–322.

129 DUBOIS, R.: Sur la perception des radiations lumineuses par la peau, chez les Protées aveugles des grottes de la Carniole. *Compt. rend. Acad. Sc.*, 1890, cx, 358–361.

130 DUBOIS, R.: Note sur l'action de la lumière sur les echinodermes (oursin). *Commun. 9me. Cong. internat. Zool., Monaco*, 1913, (1), 8–9.

131 DUSTIN, A. P.: Le rôle des tropismes et de l'odogenèse dans la régénération du système nerveux. *Arch. Biol.*, 1910, xxv, 269–388.

132 ENGELMANN, T. W.: Ueber Reizung kontraktilen Protoplasmas durch plötzliche Beleuchtung. *Arch. ges. Physiol.*, 1879, xix, 1–7.

133 ENGELMANN, T. W.: Ueber Licht- und Farbenperzeption niederster Organismen. *Arch. ges. Physiol.*, 1882, xxix, 387–400.

134 ENGELMANN, T. W.: *Bacterium photometricum*. Ein Beitrag zur vergleichenden Physiologie des Licht- und Farbensinnes. *Arch. ges. Physiol.*, 1882, xxx, 95–124.

135 ENGELMANN, T. W.: Ueber die Funktion der Otolithen. *Zool. Anz.*, 1887, x, 591, 664.

136 ENGELMANN, T. W.: Die Purpurbakterien und ihre Beziehungen zum Licht. *Bot. Ztg.*, 1888, xlvi, 661–669, 677–689, 693–701, 709–720.

137 ENGLISCH, E.: Ueber die Wirkung intermittierender Belichtungen auf Bromsilbergelatine. *Arch. wiss. Phot.*, 1899, i, 117–131.

138 ENGLISCH, E.: Ueber den zeitlichen Verlauf der durch das Licht verursachten Veränderungen der Bromsilbergelatine. *Arch. wiss. Phot.*, 1900, ii, 131–134.

139 ERHARD, H.: Beitrag zur Kenntnis des Lichtsinnes der Daphniden. *Biol. Centr.*, 1913, xxxiii, 494–496.

140 ESTERLY, C. O.: The Reactions of *Cyclops* to Light and Gravity. *Am. J. Physiol.*, 1907, xviii, 47–57.

141 EWALD, J. R.: Physiologische Untersuchungen über das Endorgan des Nervus octavus. Wiesbaden, 1892.

142 EWALD, J. R.: Ueber die Wirkung des galvanischen Stroms bei der Längsdurchströmung ganzer Wirbeltiere. *Arch. ges. Physiol.*, 1894, lv, 606–621 (Berichtigung, 1894, lvi, 354).

143 EWALD, W. E.: Ueber Orientierung, Lokomotion und Lichtreaktionen einiger Cladoceren und deren Bedeutung für die Theorie der Tropismen. *Biol. Centr.*, 1910, xxx, 1–16, 49–63, 379–399.

144 EWALD, W. E.: On Artificial Modification of Light Reactions and the Influence of Electrolytes on Phototaxis. *J. Exp. Zool.*, 1912, xiii, 591–612.

[145] EWALD, W. E.: The Applicability of the Photochemical Energy Law to Light Reactions in Animals. *Science*, 1913, xxxviii, 236–237.
[146] EWALD, W. E.: Ist die Lehre vom tierischen Phototropismus widerlegt? *Arch. Entwcklngsmech.*, 1913, xxxvii, 581–598.
[147] EWALD, W. E.: Versuche zur Analyse der Licht- und Farbenreaktionen eines Wirbellosen (*Daphnia pulex*). *Z. Sinnesphysiol.*, 1914, xlviii, 285–324.
[148] EYCLESHYMER, A. C.: The Reactions to Light of the Decapitated Young *Necturus*. *J. Comp. Neurol. and Psychol.*, 1908, xviii, 303–308.
[149] FAUVEL, P., and BOHN, G.: Le rythme des marées chez les diatomées littorales. *Compt. rend. Soc. Biol.*, 1907, lxii, 121–123.
[150] FIGDOR, W.: Ueber Helio- und Geotropismus der Gramineenblätter. *Ber. bot. Ges.*, 1905, xxiii, 182–191.
[151] FIGDOR, W.: Experimentelle Studien über die heliotropische Empfindlichkeit der Pflanzen. *Wiesner Festschrift, Wien*, 1908.
[152] FIGDOR, W.: Heliotropische Reizleitung bei Begonia-Blättern. *Ann. Jardin bot. Buitenzorg.*, 1910 (Suppl.), iii, 453–460.
[153] FIGDOR, W.: Ueber thigmotropische Empfindlichkeit der Asparagus-Sprosse. *Sitzngsb. Akad. Wiss. Wien. mathem.-naturw. Kl.* Abt. I, 1915, cxxiv, 353.
[154] FITTING, H.: Untersuchungen über den geotropischen Reizvorgang. *Jahrb. wiss. Bot.*, 1905, xli, 221–398.
[155] FLOURENS, P.: Recherches expérimentales sur les propriétés et les fonctions du système nerveux dans les animaux vertébrés. Paris, 1842, pp. xxviii + 516.
[156] FORSSMAN, J.: Ueber die Ursachen, welche die Wachstumsrichtung der peripheren Nervenfasern bei der Regeneration bestimmen. *Beitr. path. Anat.*, 1898, xxiv, 56–100.
[157] FORSSMAN, J.: Zur Kenntnis des Neurotropismus. *Beitr. path. Anat.*, 1900, xxvii, 407–430.
[158] FRANDSEN, P.: Studies on the Reactions of *Limax maximus* to Directive Stimuli. *Proc. Am. Acad. Arts and Sc.*, 1901, xxxvii, 185–227.
[159] FRANZ, V.: Phototaxis und Wanderung. Nach Versuchen mit Jungfischen und Fischlarven. *Int. Rev. ges. Hydrobiol. u. Hydrographie*, 1910, iii, 306–334.
[160] FRANZ, V.: Beiträge zur Kenntnis der Phototaxis. Nach Versuchen an Süsswassertieren. *Int. Rev. ges. Hydrobiol. u. Hydrographie, Biol. Suppl.* (2), 1911, 1–11.
[161] FRANZ, V.: Weitere Phototaxisstudien. I. Zur Phototaxis bei Fischen. II. Phototaxis bei marinen Crustaceen. III. Phototaktische Lokomotionsperioden bei *Hemimysis*. *Int. Rev. ges. Hydrobiol. u. Hydrographie, Biol. Suppl.* (3), 1911, 1–23.

LITERATURE

162 FRANZ, V.: Zur Frage der vertikalen Wanderungen der Planktontiere. *Arch. Hydrobiol. u. Planktonkunde,* 1912, vii, 493–499.

163 FRANZ, V.: Die phototaktischen Erscheinungen im Tierreiche und ihre Rolle im Freileben der Tiere. *Zool. Jahrb.,* 1913, xxxiii, 259–286.

164 v. FRISCH, K.: Ueber farbige Anpassung bei Fischen. *Zool. Jahrb.,* 1912, xxxii, 171–230.

165 v. FRISCH, K.: Sind die Fische farbenblind? *Zool. Jahrb.,* 1912, xxxiii, 107–126.

166 v. FRISCH, K.: Ueber die Farbenanpassung des *Crenilabrus*. *Zool. Jahrb.,* 1912, xxxiii, 151–164.

167 v. FRISCH, K.: Weitere Untersuchungen über den Farbensinn der Fische. *Zool. Jahrb.,* 1913, xxxiv, 43–68.

168 v. FRISCH, K.: Der Farbensinn und Formensinn der Biene. *Zool. Jahrb.,* 1914, xxxv, 1–182.

169 v. FRISCH, K., and KUPELWIESER, H.: Ueber den Einfluss der Lichtfarbe auf die phototaktischen Reaktionen niederer Krebse. *Biol. Centr.,* 1913, xxxiii, 517–552.

170 FRÖHLICH, F. W.: Vergleichende Untersuchungen über den Licht- und Farbensinn. *Deutsch. med. Wchnschr.,* 1913, xxxix, 1453–1456.

171 FRÖSCHEL, P.: Untersuchung über die heliotropische Präsentationszeit. I. *Sitzngsb. Akad. Wiss. Wien. mathem.-naturw. Kl.,* 1908, cxvii, 235–256.

172 FRÖSCHEL, P.: Untersuchung über die heliotropische Präsentationszeit. II. *Sitzngsb. Akad. Wiss. Wien. mathem.-naturw. Kl.,* 1909, cxviii, 1247–1294.

173 FUCHS, R. F.: Der Farbenwechsel und die chromatische Hautfunktion der Tiere. *Winterstein's Handb. vergl. Physiol.,* 1914, iii, I. Hälfte 2, 1189–1656.

174 GALIANO, E. F.: Beitrag zur Untersuchung der Chemotaxis der *Paramæcien*. *Z. allg. Physiol.,* 1914, xvi, 359–372.

175 GARREY, W. E.: The Effect of Ions Upon the Aggregation of Flagellated Infusoria. *Am. J. Physiol.,* 1900, iii, 291–315.

176 GARREY, W. E.: A Sight Reflex Shown by Sticklebacks. *Biol. Bull.,* 1905, viii, 79–84.

177 GARREY, W. E.: Proof of the Muscle Tension Theory of Heliotropism. *Proc. Nat. Acad. Sc.,* 1917, iii, 602–609.

178 GOLTZ, F.: Ueber die Verrichtungen des Grosshirns. I–V. *Arch. ges. Physiol.,* 1876, xiii, 1–44; 1877, xiv, 412–443; 1879, xx, 1–54; 1881, xxvi, 1–49; 1884, xxxiv, 450–505.

179 GRABER, V.: Fundamentalversuche über die Helligkeits- und Farbenempfindlichkeit augenloser und geblendeter Tiere. *Sitzngsb. Akad. Wiss. Wien,* 1883, lxxxvii, 201–236.

[180] GRABER, V.: Grundlinien zur Erforschung des Helligkeits- und Farbensinnes der Tiere. Leipzig, 1884, pp. vii +322.

[181] GRABER, V.: Ueber die Helligkeits- und Farbenempfindlichkeit einiger Meertiere. *Sitzngsb. Akad. Wiss. Wien.*, 1885, xci.

[182] GRABER, V.: Thermische Experimente an der Küchenschabe (*Periplaneta orientalis*). *Arch. ges. Physiol.*, 1887, xli, 240–256.

[183] GROOM, T. T., and LOEB, J.: Der Heliotropismus der Nauplien von *Balanus perforatus* und die periodischen Tiefenwanderungen pelagischer Tiere. *Biol. Centr.*, 1890, x, 160–177.

[184] GROSS, A. O.: The Reactions of Arthropods to Monochromatic Lights of Equal Intensities. *J. Exp. Zool.*, 1913, xiv, 467–514.

[185] HABERLANDT, G.: Ueber die Perzeption des geotropischen Reizes. *Ber. bot. Ges.*, 1900, xviii, 261–272.

[186] HADLEY, P. B.: The Relation of Optical Stimuli to Rheotaxis in the American Lobster (*Homarus americanus*). *Am. J. Physiol.*, 1906, xvii, 326–343.

[187] HADLEY, P. B.: Galvanotaxis in Larvæ of the American Lobster (*Homarus americanus*). *Am. J. Physiol.*, 1907, xix, 39–52.

[188] HADLEY, P. B.: The Reaction of Blinded Lobsters to Light. *Am. J. Physiol.*, 1908, xxi, 180–199.

[189] HADLEY, P. B.: Reactions of Young Lobsters Determined by Food Stimuli. *Science*, 1912, xxxv, 1000–1002.

[190] HARPER, E. H.: Reactions to Light and Mechanical Stimuli in the Earthworm, *Perichæta bermudensis* (Beddard). *Biol. Bull.*, 1905, x, 17–34.

[191] HARPER, E. H.: Tropic and Shock Reactions in *Perichæta* and *Lumbricus*. *J. Comp. Neurol. and Psychol.*, 1909, xix, 569–587.

[192] HARPER, E. H.: The Geotropism of *Paramæcium*. *J. Morphol.*, 1911, xxii, 993–1000.

[193] HARPER, E. H.: Magnetic Control of Geotropism in *Paramæcium*. *J. Animal Behav.*, 1912, ii, 181–189.

[194] HARRINGTON, N. R., and LEAMING, E.: The Reaction of *Amœba* to Lights of Different Colors. *Am. J. Physiol.*, 1899, iii, 9–18.

[195] HASEMAN, J. D.: The Rhythmical Movements of *Littorina littorea* Synchronous with Ocean Tides. *Biol. Bull.*, 1911, xxi, 113–121.

[196] HAUSMANN, W.: Die photodynamische Wirkung des Chlorophylls und ihre Beziehung zur photosynthetischen Assimilation der Pflanzen. *Jahrb. wiss. Bot.*, 1909, xlvi, 599–623.

[197] HELMHOLTZ, H.: Handbuch der physiologischen Optik. Hamburg, 1909–11, 3. Ed.

[198] HENRI, MME. V., and HENRI, V.: Excitation des organismes par les rayons ultra-violets. *Compt. rend. Soc. Biol.*, 1912, lxxii, 992–996; lxxiii, 326–327.

199 HENRI, V., and LARGUIER DES BANCELS, J.: Photochimie de la rétine. *J. Physiol. et Path. génér.*, 1911, xiii, 841–856.

200 HENRI, V., and LARGUIER DES BANCELS, J.: Un nouveau type de temps de réaction. *Compt. rend. Soc. Biol.*, 1912, lxxiii, 55–56.

201 HENRI, V., and LARGUIER DES BANCELS, J.: L'excitation provoquée par les rayons ultra-violets comparée avec les excitations visuelle et nerveuse, d'une part, et les réactions photochimiques, d'autre. Lois des phénomènes. *Compt. rend. Soc. Biol.*, 1912, lxxiii, 328–329.

202 HENRI, V., and LARGUIER DES BANCELS, J.: Sur l'interprétation des lois de Weber et de Jost: recherches sur les réactions des *Cyclops* exposées à la lumière ultra-violette. *Arch. Psychol.*, 1912, xii, 329–342.

203 HERBST, C.: Ueber die Bedeutung der Reizphysiologie für die kausale Auffassung von Vorgängen in der tierischen Ontogenese. I. *Biol. Centr.*, 1894, xiv, 657–666, 689–697, 727–744, 753–771, 800–810.

204 HERMANN, L.: Eine Wirkung galvanischer Ströme auf Organismen. *Arch. ges. Physiol.*, 1885, xxxvii, 457–460.

205 HERMANN, L.: Weitere Untersuchungen über das Verhalten der Froschlarven im galvanischen Strome. *Arch. ges. Physiol.*, 1886, xxxix, 414–419.

205a HERMANN, L., and MATTHIAS, F.: Der Galvanotropismus der Larven von *Rana temporaria* und der Fische. *Arch. ges. Physiol.*, 1894, lvii, 391–405.

206 HERMS, W. B.: The Photic Reactions of Sacrophagid Flies, Especially *Lucilia caesar* Linn. and *Calliphora vormitoria* Linn. *J. Exp. Zool.*, 1911, x, 167–226.

207 HERTEL, E.: Ueber die Beeinflussung des Organismus durch Licht, speciell durch die chemisch wirksamen Strahlen. *Z. allg. Physiol.*, 1904, iv, 1–43.

208 HERTEL, E.: Ueber physiologische Wirkung von Strahlen verschiedener Wellenlänge. *Z. allg. Physiol.*, 1905, v, 95–122.

209 HESS, C.: Untersuchungen über den Lichtsinn bei Fischen. *Arch. Augenheilk.*, 1909, lxiv, 1–38.

210 HESS, C.: Untersuchungen über den Lichtsinn bei wirbellosen Tieren. *Arch. Augenheilk.*, 1909, lxiv, 39–61.

211 HESS, C.: Neue Untersuchungen über den Lichtsinn bei wirbellosen Tieren. *Arch. ges. Physiol.*, 1910, cxxxvi, 282–367.

212 HESS, C.: Experimentelle Untersuchungen zur vergleichenden Physiologie des Gesichtssinnes. *Arch. ges. Physiol.*, 1911, cxlii, 405–446.

213 HESS, C.: Der Gesichtssinn. *Winterstein's Handb. vergl. Physiol.*, 1912, iv, 555–840.

214 HESS, C.: Neue Untersuchungen zur vergleichenden Physiologie des Gesichtsinnes. *Zool. Jahrb. Abt. Zool.*, 1913, xxxiii, 387–440.

215 HESS, C.: Experimentelle Untersuchungen über den angeblichen Farbensinn der Bienen. *Zool. Jahrb. Abt. Zool.*, 1913, xxxiv, 81–106.

216 HESS, C.: Eine neue Methode zur Untersuchung des Lichtsinnes bei Krebsen. *Arch. vergl. Ophthalmol.*, 1913–14, iv, 52–67.

217 HESS, C.: Untersuchungen über den Lichtsinn mariner Würmer und Krebse. *Arch. ges. Physiol.*, 1914, clv, 421–435.

218 HESS, C.: Untersuchungen über den Lichtsinn bei Echinodermen. *Arch. ges. Physiol.*, 1914, clx, 1–26.

219 HESS, C.: Messende Untersuchung des Lichtsinnes der Biene. *Arch. ges. Physiol.*, 1916, clxiii, 289–320.

220 HESSE, R.: Untersuchungen über die Organe der Lichtempfindung bei niederen Tieren. I. Die Organe der Lichtempfindung bei den Lubriciden. *Z. wiss. Zool.*, 1896, lxi, 393–419.

221 HESSE, R.: II. Die Augen der Plathelminthen, insonderheit der trikladen Turbellarien. *Z. wiss. Zool.*, 1897, lxii, 527–582.

222 HESSE, R.: IV. Die Sehorgane des *Amphioxus*. *Z. wiss. Zool.*, 1898, lxiii, 456–464.

223 HESSE, R.: Die Lichtempfindung des *Amphioxus*. *Anat. Anz.*, 1898, xiv, 556.

224 HÖGYES, A.: Der Nervenmechanismus der assoziierten Augenbewegungen. I–II. *Mitt. mathem.-naturw. Kl. Ungar. Akad. Wiss. Budapest*, 1881, x, 1–62; xi, 1–100. (Ref. *Biol. Centr.*, 1881–82, i, 216–220.)

225 HOLMES, S. J.: Phototaxis in the Amphipoda. *Am. J. Physiol.*, 1901, v, 211–234.

226 HOLMES, S. J.: Phototaxis in *Volvox*. *Biol. Bull.*, 1903, iv, 319–326.

227 HOLMES, S. J.: The Selection of Random Movements as a Factor in Phototaxis. *J. Comp. Neurol. and Psychol.*, 1905, xv, 98–112.

228 HOLMES, S. J.: The Reactions of *Ranatra* to Light. *J. Comp. Neurol. and Psychol.*, 1905, xv, 305–349.

229 HOLMES, S. J.: Observations on the Young of *Ranatra quadridentata* Stal. *Biol. Bull.*, 1907, xii, 158–164.

230 HOLMES, S. J.: Phototaxis in Fiddler Crabs and Its Relation to Theories of Orientation. *J. Comp. Neurol. and Psychol.*, 1908, xviii, 493–497.

231 HOLMES, S. J.: Description of a New Species of *Eubranchipus* from Wisconsin with Observations on Its Reaction to Light. *Trans. Wis. Acad. Sc., Arts and Letters*, 1910, xvi, pt. II, 1252–1255.

232 HOLMES, S. J.: Pleasure, Pain and the Beginnings of Intelligence. *J. Comp. Neurol. and Psychol.*, 1910, xx, 145–164.

233 HOLMES, S. J.: Evolution of Animal Intelligence. New York, 1911, pp. 296.

LITERATURE

234 HOLMES, S. J.: The Reactions of Mosquitoes to Light in Different Periods of Their History. *J. Animal Behav.*, 1911, i, 29–32.

235 HOLMES, S. J.: The Beginnings of Intelligence. *Science*, 1911, xxxiii, 473–480.

236 HOLMES, S. J.: The Tropisms and Their Relation to More Complex Modes of Behavior. *Bull. Wis. Nat. Hist. Soc.*, 1912, x, 13–23.

237 HOLMES, S. J.: Phototaxis in the Sea Urchin, *Arbacia punctulata*. *J. Animal Behav.*, 1912, ii, 126–136.

238 HOLMES, S. J.: Studies in Animal Behavior. Boston, 1916, pp. 266.

239 HOLMES, S. J., and MCGRAW, K. W.: Some Experiments on the Method of Orientation to Light. *J. Animal Behav.*, 1913, iii, 367–373.

240 HOLT, E. B., and LEE, F. S.: The Theory of Phototactic Response. *Am. J. Physiol.*, 1901, iv, 460–481.

241 HOWARD, L. O.: Butterflies Attracted to Light at Night. *Proc. Ent. Soc., Washington*, 1889, iv.

242 ILYIN, P.: Das Gehörbläschen als Gleichgewichtsorgan bei den Pterotracheidæ. *Centr. Physiol.*, 1900, xiii, 691–694.

243 JACKSON, H. H. T.: The Control of Phototactic Reactions in *Hyalella* by Chemicals. *J. Comp. Neurol. and Psychol.*, 1910, xx, 259–263.

244 JENNINGS, H. S.: Studies on Reactions to Stimuli in Unicellular Organisms. I. Reactions to Chemical, Osmotic and Mechanical Stimuli in the Ciliate Infusoria. *J. Physiol.*, 1897, xxi, 258–322.

245 JENNINGS, H. S.: II. The Mechanism of the Motor Reactions of *Paramæcium*. *Am. J. Physiol.*, 1899, ii, 311–341.

246 JENNINGS, H. S.: III. Reactions to Localized Stimuli in *Spirostomum* and *Stentor*. *Am. Nat.*, 1899, xxxiii, 373–389.

247 JENNINGS, H. S.: V. On the Movements and Motor Reflexes of the Flagellata and Ciliata. *Am. J. Physiol.*, 1900, iii, 229–260.

248 JENNINGS, H. S.: VI. On the Reactions of *Chilomonas* to Organic Acids. *Am. J. Physiol.*, 1900, iii, 397–403.

249 JENNINGS, H. S., and CROSBY, J. H.: VII. The Manner in Which Bacteria React to Stimuli, Especially to Chemical Stimuli. *Am. J. Physiol.*, 1901, vi, 31–37.

250 JENNINGS, H. S., and MOORE, E. H.: VIII. On the Reactions of Infusoria to Carbonic and Other Acids, with Especial Reference to the Causes of the Gatherings Spontaneously Formed. *Am. J. Physiol.*, 1902, vi, 233–250.

251 JENNINGS, H. S.: Contributions to the Study of the Behavior of Lower Organisms. *Carnegie Institution of Washington*, Pub. No. 16, 1904, pp. 256, 81 figs.

252 JENNINGS, H. S.: Modifiability in Behavior. II. Factors Determining Direction and Character of Movement in the Earthworm. *J. Exp. Zool.*, 1906, iii, 435–455.

253 JENNINGS, H. S.: Behavior of the Lower Organisms. New York, 1906, pp. xiv + 366.

254 JENNINGS, H. S.: The Interpretation of the Behavior of the Lower Organisms. *Science*, 1908, xxvii, 698–710.

255 JENNINGS, H. S.: Tropisms. *Rapport VIme Congr. Internat. Psychol. Genève*, 1909, pp. 20.

255a JENSEN, P.: Ueber den Geotropismus niederer Organismen. *Arch. ges. Physiol.*, 1893, liii, 428–480.

256 JORDAN, H.: Rheotropic Responses of *Epinephelus striatus* Bloch. *Am. J. Physiol.*, 1917, xliii, 438–454.

257 JORDAN, H.: Integumentary Photosensitivity in a Marine Fish, *Epinephelus striatus* Bloch. *Am. J. Physiol.*, 1917, xliv, 259–274.

258 JUDAY, C.: The Diurnal Movement of Plancton Crustacea. *Trans. Wis. Acad. Sc., Arts and Letters*, 1904, xiv, 534–568.

259 KAFKA, G.: Einführung in die Tierpsychologie. I. Die Sinne der Wirbellosen. Leipzig, 1914, xii+594.

260 KANDA, S.: On the Geotropism of *Paramœcium* and *Spirostomum*. *Biol. Bull.*, 1914, xxvi, 1–24.

261 KANDA, S.: The Reversibility of the Geotropism of *Arenicola* Larvæ by Salts. *Am. J. Physiol.*, 1914, xxxv, 162–176.

262 KANDA, S.: Geotropism in Animals. *Am. J. Psychol.*, 1915, xxvi, 417–427.

263 KANDA, S.: Studies on the Geotropism of the Marine Snail, *Littorina littorea*. *Biol. Bull.*, 1916, xxx, 57–84.

264 KANDA, S.: The Geotropism of Freshwater Snails. *Biol. Bull.*, 1916, xxx, 85–97.

264a KANDA, S.: Further Studies on the Geotropism of *Paramœcium caudatum*. *Biol. Bull.*, 1918, xxxiv, 108–119.

265 KELLOGG, V. L.: Some Insect Reflexes. *Science*, 1903, xviii, 693–696.

266 KELLOGG, V. L.: Some Silkworm Moth Reflexes. *Biol. Bull.*, 1907, xii, 152–154.

267 KNIEP, H.: Untersuchungen über die Chemotaxis von Bakterien. *Jahrb. wiss. Bot.*, 1906, xliii, 215–270.

268 KRANICHFELD, H.: Zum Farbensinn der Bienen. *Biol. Centr.*, 1915, xxxv, 39–46.

269 KRECKER, F. H.: Phenomena of Orientation Exhibited by Ephemeridæ. *Biol. Bull.*, 1915, xxix, 381–388.

270 KREIDL, A.: Weitere Beiträge zur Physiologie des Ohrlabyrinthes. *Sitzngsb. Akad. Wiss. Wien, mathem.-naturw. Kl.*, 1892, ci, 469–480; 1893, cii, 149–174.

LITERATURE

[271] KRIBS, H. G.: The Reactions of *Æolosoma* to Chemical Stimuli. *J. Exp. Zool.*, 1910, viii, 43–74.

[272] KÜHNE, W.: Untersuchungen über das Protoplasma und die Kontractilität. Leipzig, 1864, pp. 158.

[273] KÜHNE, W.: Chemische Vorgänge in der Netzhaut. *Hermann's Handb. Physiol.*, 1879, iii, pt. 1, 235–342.

[274] KUSANO, S.: Studies on the Chemotactic and Other Related Reactions of the Swarmspores of Myxomycetes. *J. Coll. Agriculture, Imp. Univ. Tokyo*, 1909, ii.

[275] LASAREFF, P.: Ionentheorie der Nerven- und Muskelreizung. *Arch. ges. Physiol.*, 1910, cxxxv, 196–204.

[276] LAURENS, H.: The Reactions of Amphibians to Monochromatic Lights of Equal Intensity. *Bull. Mus. Comp. Zool.*, 1911, xliii, 253–302.

[277] LAURENS, H.: The Reactions of Normal and Eyeless Amphibian Larvæ to Light. *J. Exp. Zool.*, 1914, xvi, 195–210.

[278] LEE, F. S.: A Study of the Sense of Equilibrium in Fishes. I. *J. Physiol.*, 1893, xv, 311–348.

[279] LEE, F. S.: A Study of the Sense of Equilibrium in Fishes. II. *J. Physiol.*, 1894–95, xvii, 192–210.

[280] LIDFORSS, B.: Ueber den Chemotropismus der Pollenschläuche. *Ber. bot. Ges.*, 1899, xvii, 236–242.

[281] LIDFORSS, B.: Ueber die Reizbewegungen der *Marchantia*-Spermatozoiden. *Jahrb. wiss. Bot.*, 1905, xli, 65–87.

[282] LIDFORSS, B.: Ueber die Chemotaxis eines Thiospirillum. *Ber. bot. Ges.*, 1912, xxx, 262–274.

[283] LILLIE, F. R.: Studies of Fertilization. V. The Behavior of the Spermatozoa of *Nereis* and *Arbacia* with Special Reference to Egg-extractives. *J. Exp. Zool.*, 1913, xiv, 515–574.

[284] LOEB, J.: Beiträge zur Physiologie des Grosshirns. *Arch. ges. Physiol.*, 1886, xxxix, 265–346.

[285] LOEB, J.: Die Orientierung der Tiere gegen das Licht (tierischer Heliotropismus). *Sitzngsb. Würzb. physik.-med. Ges.*, 1888.

[286] LOEB, J.: Die Orientierung der Tiere gegen die Schwerkraft der Erde (tierischer Geotropismus). *Sitzngsb. Würzb. physik.-med. Ges.*, 1888.

[287] LOEB, J.: Der Heliotropismus der Tiere und seine Uebereinstimmung mit dem Heliotropismus der Pflanzen. Würzburg, 1889, pp. 118.

[288] LOEB, J.: Weitere Untersuchungen über den Heliotropismus der Tiere und seine Uebereinstimmung mit dem Heliotropismus der Pflanzen. (Heliotropische Krümmungen bei Tieren). *Arch. ges. Physiol.*, 1890, xlvii, 391–416.

[289] LOEB, J.: Ueber Geotropismus bei Tieren. *Arch. ges. Physiol.*, 1891, xlix, 175–189.

[290] LOEB, J.: Ueber den Anteil des Hörnerven an den nach Gehirnverletzung auftretenden Zwangsbewegungen, Zwangslagen und assoziierten Stellungsänderungen der Bulbi und Extremitäten. *Arch. ges. Physiol.*, 1891, l, 66–83.

[290a] LOEB, J.: Untersuchungen zur physiologischen Morphologie der Tiere. I. Heteromorphose. II. Organbildung und Wachstum. Würzburg, 1891–1892.

[291] LOEB, J.: Ueber künstliche Umwandlung positiv heliotropischer Tiere in negativ heliotropische und umgekehrt. *Arch. ges. Physiol.*, 1893, liv, 81–107.

[292] LOEB, J.: Zur Theorie der physiologischen Licht- und Schwerkraftwirkungen. *Arch. ges. Physiol.*, 1897, lxvi, 439–466.

[293] LOEB, J.: Comparative Physiology of the Brain and Comparative Psychology. New York, 1900, x+309.

[294] LOEB, J.: Studies in General Physiology. Chicago, 1905, 2 vols., x+782.

[295] LOEB, J.: The Dynamics of Living Matter. New York, 1906, xi+233.

[296] LOEB, J.: Ueber die Erregung von positivem Heliotropismus durch Säure, insbesondere Kohlensäure und von negativem Heliotropismus durch ultraviolette Strahlen. *Arch. ges. Physiol.*, 1906, cxv, 564–581.

[297] LOEB, J.: Concerning the Theory of Tropisms. *J. Exp. Zool.*, 1907, iv, 151–156.

[298] LOEB, J.: Ueber die Summation heliotropischer und geotropischer Wirkungen bei den auf der Drehscheibe ausgelösten kompensatorischen Kopfbewegungen. *Arch. ges. Physiol.*, 1907, cxvi, 368–374.

[299] LOEB, J.: Chemische Konstitution und physiologische Wirksamkeit von Alkoholen und Säuren. II. *Biochem. Z.*, 1909, xxiii, 93–96.

[300] LOEB, J.: Die Tropismen. *Winterstein's Handb. vergl. Physiol.*, 1911, iv, 451–519.

[301] LOEB, J.: The Mechanistic Conception of Life. Chicago, 1912, pp. 232.

[302] LOEB, J.: On the Nature of the Conditions Which Determine or Prevent the Entrance of the Spermatozoon into the Egg. *Am. Nat.*, 1915, xlix, 257–285.

[303] LOEB, J.: The Organism as a Whole. From a Physico-chemical Viewpoint. New York, 1916, pp. 379.

303a LOEB, J.: The Chemical Basis of Regeneration and Geotropism. *Science*, 1917, xlvi, 115–118.

303b LOEB, J.: Influence of the Leaf upon Root Formation and Geotropic Curvature in the Stem of *Bryophyllum calycinum* and the Possibility of a Hormone Theory of These Processes. *Bot. Gaz.*, 1917, lxiii, 25–50.

303c LOEB, J.: The Chemical Mechanism of Regeneration. *Ann. Inst. Pasteur*, 1918, xxxii, 1–16.

304 LOEB, J., and BUDGETT, S. P.: Zur Theorie des Galvanotropismus. IV. Ueber die Ausscheidung electropositiver Ionen an der äusseren Anodenfläche protoplasmatischer Gebilde als Ursache der Abweichungen vom Pflüger'schen Erregungsgesetz. *Arch. ges. Physiol.*, 1897, lxv, 518–534.

305 LOEB, J., and EWALD, W. F.: Ueber die Gültigkeit des Bunsen-Roscoe'schen Gesetzes für die heliotropische Erscheinung bei Tieren. *Centr. Physiol.*, 1914, xxvii, 1165–1168.

306 LOEB, J., and GARREY, W. E.: Zur Theorie des Galvanotropismus. II. Versuche an Wirbeltieren. *Arch. ges. Physiol.*, 1896, lxv, 41–47.

307 LOEB, J., and MAXWELL, S. S.: Zur Theorie des Galvanotropismus. *Arch. ges. Physiol.*, 1896, lxiii, 121–144.

308 LOEB, J., and MAXWELL, S. S.: Further Proof of the Identity of Heliotropism in Animals and Plants. *Univ. Cal. Pub. Physiol.*, 1910, iii, 195–197.

309 LOEB, J., and NORTHROP, J. H.: Heliotropic Animals as Photometers on the Basis of the Validity of the Bunsen-Roscoe Law for Heliotropic Reactions. *Proc. Nat. Acad. Sc.*, 1917, iii, 539–544.

310 LOEB, J., and WASTENEYS, H.: On the Identity of Heliotropism in Animals and Plants. *Proc. Nat. Acad. Sc.*, 1915, i, 44–47; *Science*, 1915, xli, 328–330.

311 LOEB, J., and WASTENEYS, H.: The Relative Efficiency of Various Parts of the Spectrum for the Heliotropic Reactions of Animals and Plants. *J. Exp. Zool.*, 1915, xix, 23–35; 1916, xx, 217–236.

312 LOEB, J., and WASTENEYS, H.: A Re-examination of the Applicability of the Bunsen-Roscoe Law to the Phenomena of Animal Heliotropism. *J. Exp. Zool.*, 1917, xxii, 187–192.

313 LÖHNER, L.: Untersuchungen über den sogenannten Totstellreflex der Arthropoden. *Z. allg. Physiol.*, 1914, xvi, 373–418.

314 LUBBOCK, J.: On the Sense of Color Among Some of the Lower Animals. I and II. *J. Linn. Soc. (Zool.)*, 1881, xvi, 121–127; 1882, xvii, 205–214.

315 LUBBOCK, J.: On the Senses, Instincts and Intelligence of Animals, with Special Reference to Insects. Internat. sc. Series, London, 1899.

[316] Lubbock, J.: Ants, Bees and Wasps. New York, 1904, xiii+435.
[317] Ludloff, K.: Untersuchungen über den Galvanotropismus. *Arch. ges. Physiol.*, 1895, lix, 525–554.
[318] Lyon, E. P.: The Functions of the Otocyst. *J. Comp. Neurol. and Psychol.*, 1898, viii, 238–245.
[319] Lyon, E. P.: A Contribution to the Comparative Physiology of Compensatory Motions. *Am. J. Physiol.*, 1899, iii, 86–114.
[320] Lyon, E. P.: Compensatory Motions in Fishes. *Am. J. Physiol.*, 1900, iv, 77–82.
[321] Lyon, E. P.: On Rheotropism. I. Rheotropism in Fishes. *Am. J. Physiol.*, 1904, xii, 149–161.
[322] Lyon, E. P.: Rheotropism in Fishes. *Biol. Bull.*, 1905, viii, 238–239.
[323] Lyon, E. P.: On the Theory of Geotropism in *Paramæcium*. *Am. J. Physiol.*, 1905, xiv, 421–432.
[324] Lyon, E. P.: Note on the Geotropism of *Arbacia* Larvæ. *Biol. Bull.*, 1906, xii, 21–22.
[325] Lyon, E. P.: Note on the Heliotropism of *Palæmonetes* Larvæ. *Biol. Bull.*, 1906, xii, 23–25.
[326] Lyon, E. P.: On Rheotropism. II. Rheotropism of Fish Blind in One Eye. *Am. J. Physiol.*, 1909, xxiv, 244–251.
[326a] Lyon, E. P.: Note on the Geotropism of *Paramæcium*. *Biol. Bull.*, 1918, xxxiv, 120.
[326b] McClendon, J. F.: Protozoan Studies. *J. Exp. Zool.*, 1909, vi, 265–283.
[327] MacCurdy, H.: Some Effects of Sunlight in the Starfish. *Science*, 1913, xxxvi, 98–100.
[327a] McEwen, R. S.: The Reactions to Light and to Gravity in *Drosophila* and its Mutants. *J. Exp. Zool.*, 1918, xxv, 49–106.
[328] McGinnis, M. O.: Reactions of *Branchipus serratus* to Light, Heat and Gravity. *J. Exp. Zool.*, 1911, x, 227–240.
[329] Mach, E.: Physikalische Versuche über den Gleichgewichtssinn des Menschen. *Sitzngsb. Akad. Wiss. Wien.*, 1873, lxviii; 1874, lxix.
[330] Mach, E.: Grundlinien der Lehre von den Bewegungsempfindungen. Leipzig, 1875, pp. 127.
[331] Mach, E.: Beiträge zur Analyse der Empfindungen. Jena, 1902.
[332] Magnus, R.: Welche Teile des Zentralnervensystems müssen für das Zustandekommen der tonischen Hals- und Labyrinthreflexe auf die Körpermuskulatur vorhanden sein? *Arch. ges. Physiol.*, 1914, clix, 224–250.
[333] Magnus, R., and De Kleijn, A.: Die Abhängigkeit des Tonus der Extremitätenmuskeln von der Kopfstellung. *Arch. ges. Physiol.*, 1912, cxlv, 455–548.

334 MAGNUS, R., and DE KLEIJN, A.: Die Abhängigkeit des Tonus der Nackenmuskeln von der Kopfstellung. *Arch. ges. Physiol.*, 1912, cxlvii, 403–416.

335 MAGNUS, R., and DE KLEIJN, A.: Die Abhängigkeit der Körperstellung vom Kopfstande beim normalen Kaninchen. *Arch. ges. Physiol.*, 1913, cliv, 163–177.

336 MAGNUS, R., and DE KLEIJN, A.: Analyse der Folgezustände einseitiger Labyrinthexstirpation mit besonderer Berücksichtigung der Rolle der tonischen Halsreflexe. *Arch. ges. Physiol.*, 1913, cliv, 178–306.

337 MAGNUS, R., and VAN LEEUWEN, W. S.: Die akuten und die dauernden Folgen des Ausfalles der tonischen Hals- und Labyrinthreflexe. *Arch. ges. Physiol.*, 1914, clix, 157–217.

338 MAGNUS, R., and WOLF, C. G. L.: Weitere Mitteilungen über den Einfluss der Kopfstellung auf den Gliedertonus. *Arch. ges. Physiol.*, 1913, cxlix, 447–461.

339 MARCHAL, P.: Le retour au nid chez le *Pompilus sericeus* V. d. L. *Compt. rend. Soc. Biol.*, 1900, lii, 1113–1115.

340 MASSART, J.: Recherches sur les organismes inférieurs. I. La loi du Weber vérifiée pour l'héliotropisme du champignon. *Bull. Acad. Roy. Belg.*, 1888, (3) xvi, 590.

341 MASSART, J.: Sur l'irritabilité des spermatozoides dans l'oeuf de la grenouille. *Bull. Acad. Roy. Belg.*, 1888, (3) xv; 1889, xviii.

342 MASSART, J.: La sensibilité tactile chez les organismes inférieurs. *J. Soc. Roy. Sc. med. et nat., Bruxelles*, 1890.

343 MASSART, J.: Recherches sur les organismes inférieurs. III. La sensibilité à la gravitation. *Bull. Acad. Roy. Belg.*, 1891, (3) xxii, 158–167.

344 MASSART, J.: Essai de classification des réflexes non-nerveux. *Ann. Inst. Pasteur*, 1901, xv, 635–672.

345 MASSART, J.: Versuch einer Einteilung der nichtnervösen Reflexe. *Biol. Centr.*, 1902, xxii, 9–23.

346 MAST, S. O.: Light and the Behavior of Organisms. New York, 1911, pp. 410+xi.

347 MAST, S. O.: Behavior of Fire-flies (*Photinus pyralis?*) with Special Reference to the Problem of Orientation. *J. Animal Behav.*, 1912, ii, 256–272.

348 MAST, S. O.: The Relation between Spectral Color and Stimulation in the Lower Organisms. *J. Exp. Zool.*, 1917, xxii, 471–528.

348a MATULA, J.: Untersuchungen über die Funktionen des Zentralnervensystems bei Insekten. *Arch. ges. Physiol.*, 1911, cxxxviii, 388–456.

349 MAXWELL, S. S.: Beiträge zur Gehirnphysiologie der Anneliden. *Arch. ges. Physiol.*, 1897, lxvii, 263–297.

350 MAXWELL, S. S.: Experiments on the Functions of the Internal Ear. *Univ. Cal. Pub. Physiol.*, 1910, iv, 1–4.

351 MAYER, A. G., and SOULE, C. G.: Some Reactions of Caterpillars and Moths. *J. Exp. Zool.*, 1906, iii, 415–433.

352 MENDELSSOHN, M.: Ueber den Thermotropismus einzelliger Organismen. *Arch. ges. Physiol.*, 1895, lx, 1–27.

353 MENDELSSOHN, M.: Recherches sur la thermotaxie des organismes unicellulaires. *J. Physiol. et Path. génér.*, 1902, iv, 393–409.

354 MENDELSSOHN, M.: Recherches sur l'interférence de la thermotaxie avec d'autres tactismes et sur le mécanisme du mouvement thermotactique. *J. Physiol. et Path. génér.*, 1902, iv, 475–488.

355 MENDELSSOHN, M.: Quelques considérations sur la nature et le rôle biologique de la thermotaxie. *J. Physiol. et Path. génér.*, 1902, iv, 489–496.

356 MENKE, H.: Periodische Bewegungen und ihr Zusammenhang mit Licht und Stoffwechsel. *Arch. ges. Physiol.*, 1911, cxl, 37–91.

357 MEREJKOWSKY, C. DE: Les crustacés inférieurs distinguent-ils les couleurs? *Compt. rend. Acad. Sc.*, 1881, xciii, 1160–1161.

358 MILLER, F. R.: Galvanotropism in the Crayfish. *J. Physiol.*, 1907, xxxv, 215–229.

359 MINKIEWICZ, R.: Sur le chromotropisme et son inversion artificielle. *Compt. rend. Acad. Sc.*, 1906, cxliii, 785–787.

360 MINKIEWICZ, R.: Le rôle des phénomènes chromotropiques dans l'étude des problèmes biologiques et psycho-physiologiques. *Compt. rend. Acad. Sc.*, 1906, cxliii, 934–935.

361 MINKIEWICZ, R.: Une expérience sur la nature du chromotropisme chez les némertes. *Compt. rend. Acad. Sc.*, 1912, clv, 229–231.

362 MITSUKURI, K.: Negative Phototaxis and Other Properties of *Littorina* as Factors in Determining Its Habitat. *Annotationes Zoologicæ Japonenses*, 1901, iv, 1–19.

363 MOLISCH, H.: Untersuchungen über den Hydrotropismus. *Sitzngsb. Akad. Wiss. Wien. mathem.-naturw. Kl.*, 1883.

364 MOORE, ANNE: Some Facts Concerning Geotropic Gatherings of *Paramæcia*. *Am. J. Physiol.*, 1903, ix, 238–244.

365 MOORE, A. R.: On the Righting Movements of the Starfish. *Biol. Bull.*, 1910, xix, 235–239.

366 MOORE, A. R.: Concerning Negative Phototropism in *Daphnia pulex*. *J. Exp. Zool.*, 1912, xiii, 573–575.

367 MOORE, A. R.: Negative Phototropism in *Diaptomus* by Means of Strychnine. *Univ. Cal. Pub. Physiol.*, 1912, iv, 185–186.

368 MOORE, A. R.: The Negative Phototropism of *Diaptomus* Through the Agency of Caffein, Strychnine, and Atropin. *Science*, 1913, xxxviii, 131–133.

369 MOORE, A. R.: The Mechanism of Orientation in *Gonium*. *J. Exp. Zool.*, 1916, xxi, 431–432.

369a MOORE, A. R.: The Action of Strychnine on Certain Invertebrates. *J. Pharm. and Exp. Therap.*, 1916, ix, 167–169.

370 MOORE, A. R., and KELLOGG, F. M.: Note on the Galvanotropic Response of the Earthworm. *Biol. Bull.*, 1916, xxx, 131–134.

371 MOORE, B.: Observations of Certain Marine Organisms of (a) Variations in Reaction to Light, and (b) a Diurnal Periodicity of Phosphorescence. *Biochem. J.*, 1909, iv, 1–29.

371a MORGAN, C. L.: Animal Behavior. London, 1900.

371b MORGULIS, S.: The Auditory Reactions of the Dog Studied by the Pawlow Method. *J. Animal Behav.*, 1914, iv, 142–145.

371c MORGULIS, S.: Pawlow's Theory of the Function of the Central Nervous System and a Digest of Some of the More Recent Contributions to This Subject from Pawlow's Laboratory. *J. Animal Behav.*, 1914, iv, 362–379.

372 MORSE, M. W.: Alleged Rhythm in Phototaxis Synchronous with Ocean Tides. *Proc. Soc. Exp. Biol. and Med.*, 1910, vii, 145–146.

373 MÜLLER, H.: Ueber Heliotropismus. *Flora*, 1876, lix, 65–70, 88–95.

374 MÜLLER-HETTLINGEN, J.: Ueber galvanische Erscheinungen an keimenden Samen. *Arch. ges. Physiol.*, 1883, xxxi, 193–212.

375 MURBACH, L.: The Static Function in *Gonionemus*. *Am. J. Physiol.*, 1903, x, 201–209.

376 MURBACH, L.: Some Light Reactions of the Medusa *Gonionemus*. *Biol. Bull.*, 1909, xvii, 354–368.

377 MUSSET, CH.: Sélénotropisme. *Compt. rend. Acad. Sc.*, 1890, cx, 201–202.

378 NAGEL, W. A.: Beobachtungen über den Lichtsinn augenloser Muscheln. *Biol. Centr.*, 1894, xiv, 385–390.

379 NAGEL, W. A.: Ein Beitrag zur Kenntnis des Lichtsinnes augenloser Tiere. *Biol. Centr.*, 1894, xiv, 810–813.

379a NAGEL, W. A.: Experimentelle sinnesphysiologische Untersuchungen an Cœlenteraten. *Arch. ges. Physiol.*, 1894, lvii, 495–552.

380 NAGEL, W. A.: Ueber Galvanotaxis. *Arch. ges. Physiol.*, 1895, lix, 603–642.

381 NAGEL, W. A.: Der Lichtsinn augenloser Tiere. Jena, 1896, pp. 120.

382 NAGEL, W. A.: Phototaxis, Photokinesis und Unterschiedsempfindlichkeit. *Bot. Ztg.*, 1901, lix, 298–299.

[383] NAGEL, W. A.: Methoden zur Erforschung des Licht- und Farbensinnes. *Tigerstedt's Handb. physiol. Methodik*, 1909, iii, Abt. 2, Sinnesphysiologie, ii, 1–99.

[384] NATHANSOHN, A., and PRINGSHEIM, E.: Ueber die Summation intermittierender Lichtreize. *Jahrb. wiss. Bot.*, 1908, xlv, 137–190.

[385] NĚMEC, B.: Ueber die Wahrnehmung des Schwerkraftreizes bei den Pflanzen. *Jahrb. wiss. Bot.*, 1901, xxxvi, 80–178.

[385a] NERNST, W., and BARRATT, J. O. W.: Ueber die elektrische Nervenreizung durch Wechselströme. *Z. Electrochem.*, 1904, x, 664–668.

[386] NEUBERG, C.: Chemische Umwandlungen durch Strahlenarten. *Biochem. Z.*, 1908, xiii, 305–320; 1909, xvii, 270–292.

[387] NUEL, J. P.: La vision. Paris, 1904, pp. 376.

[388] NYBERGH, T.: Studien über die Einwirkung der Temperatur auf die tropistische Reisbarkeit etiolierter *Avena*-Keimlinge. *Ber. bot. Ges.*, 1912, xxx, 542–553.

[389] OLTMANNS, F.: Ueber die photometrischen Bewegungen der Pflanzen. *Flora*, 1892, lxxv, 183–266.

[390] OLTMANNS, F.: Ueber positiven und negativen Heliotropismus. *Flora*, 1897, lxxxiii, 1.

[391] OSTWALD, WO.: Ueber eine neue theoretische Betrachtungsweise in der Planktologie, insbesondere über die Bedeutung des Begriffs der "inneren Reibung des Wassers" für dieselbe. *Forsch.-ber. biol. Station Plön*, 1903, pt. 10, 1–49.

[392] OSTWALD, WO.: Zur Theorie der Richtungsbewegungen schwimmender niederer Organismen. *Arch. ges. Physiol.*, 1903, xcv, 23–65; 1906, cxi, 452–472; 1907, cxvii, 384–408.

[393] OSTWALD, WO.: Ueber die Lichtempfindlichkeit tierischer Oxydasen und über die Beziehungen dieser Eigenschaft zu den Erscheinungen des tierischen Phototropismus. *Biochem. Z.*, 1908, x, 1–130.

[394] PAAL, A.: Ueber phototropische Reizleitungen. *Ber. bot. Ges.*, 1914, xxxii, 499–502.

[395] PARKER, G. H.: Photomechanical Changes in the Retinal Pigment Cells of *Palæmonetes,* and Their Relation to the Central Nervous System. *Bull. Mus. Comp. Zool.*, 1897, xxx, 273–300.

[396] PARKER, G. H.: The Photomechanical Changes in the Retinal Pigment of *Gammarus*. *Bull. Mus. Comp. Zool.*, 1899, xxxv, 141–148.

[397] PARKER, G. H.: The Reactions of Copepods to Various Stimuli and the Bearing of This on Daily Depth-migrations. *Bull. U. S. Fish Comm.*, 1901, 103–123.

[398] PARKER, G. H.: The Phototropism of the Mourning-cloak Butterfly, *Vanessa antiopa* Linn. *Mark Anniversary Vol.*, 1903, 453–469.

399 PARKER, G. H.: The Skin and the Eyes as Receptive Organs in the Reactions of Frogs to Light. *Am. J. Physiol.*, 1903, x, 28–36.

400 PARKER, G. H.: The Stimulation of the Integumentary Nerves of Fishes by Light. *Am. J. Physiol.*, 1905, xiv, 413–420.

401 PARKER, G. H.: The Reactions of *Amphioxus* to Light. *Proc. Soc. Exp. Biol. and Med.*, 1906, iii, 61–62.

402 PARKER, G. H.: The Influence of Light and Heat on the Movement of the Melanophore Pigment, Especially in Lizards. *J. Exp. Zool.*, 1906, iii, 401–414.

403 PARKER, G. H.: The Sensory Reactions of *Amphioxus*. *Proc. Am. Acad. Arts and Sc.*, 1908, xliii, 415–455.

404 PARKER, G. H.: The Integumentary Nerves of Fishes as Photoreceptors and Their Significance for the Origin of the Vertebrate Eyes. *Am. J. Physiol.*, 1909, xxv, 77–80.

405 PARKER, G. H.: Mast's "Light and the Behavior of Organisms." *J. Animal Behav.*, 1911, i, 461–464.

406 PARKER, G. H., and ARKIN, L.: The Directive Influence of Light on the Earthworm *Allolobophora fœtida* (Sav.). *Am. J. Physiol.*, 1901, v, 151–157.

407 PARKER, G. H., and BURNETT, F. L.: The Reactions of *Planarians*, With and Without Eyes, to Light. *Am. J. Physiol.*, 1900, iv, 373–385.

408 PARKER, G. H., and METCALF, C. R.: The Reactions of Earthworms to Salts: a Study in Protoplasmic Stimulation as a Basis of Interpreting the Sense of Taste. *Am. J. Physiol.*, 1906, xvii, 55–74.

409 PARKER, G. H., and PARSHLEY, H. M.: The Reactions of Earthworms to Dry and to Moist Surfaces. *J. Exp. Zool.*, 1911, xi, 361–363.

410 PARKER, G. H., and PATTEN, B. M.: The Physiological Effect of Intermittent and of Continuous Lights of Equal Intensities. *Am. J. Physiol.*, 1912, xxxi, 22–29.

411 PARMLEE, M.: The Science of Human Behavior. New York, 1913, xvii+443.

412 PATTEN, B. M.: A Quantitative Determination of the Orienting Reaction of the Blowfly Larva (*Calliphora erythrocephala* Meigen), *J. Exp. Zool.*, 1914, xvii, 213–280.

413 PATTEN, B. M.: An Analysis of Certain Photic Reactions with Reference to the Weber-Fechner Law. I. The Reactions of the Blowfly Larva to Opposed Beams of Light. *Am. J. Physiol.*, 1915, xxxviii, 313–338.

414 PATTEN, B. M.: The Changes of the Blowfly Larva's Photosensitivity with Age. *J. Exp. Zool.*, 1916, xx, 585–598.

415 PATTEN, B. M.: Reactions of the Whip-tail Scorpion to Light. *J. Exp. Zool.*, 1917, xxiii, 251–275.
416 PAYNE, F.: The Reactions of the Blind Fish, *Amblyopsis spelæus*, to Light. *Biol. Bull.*, 1907, xiii, 317–323.
416a PAYNE, F.: Forty-nine Generations in the Dark. *Biol. Bull.*, 1910, xviii, 188–190.
416b PAYNE, F.: *Drosophila ampelophila* Loew Bred in the Dark for Sixty-nine Generations. *Biol. Bull.*, 1911, xxi, 297–301.
417 PEARL, R.: Studies on Electrotaxis. I. On the Reactions of Certain Infusoria to the Electric Current. *Am. J. Physiol.*, 1900, iv, 96–123.
418 PEARL, R.: Studies on the Effects of Electricity on Organisms. II. The Reactions of *Hydra* to the Constant Current. *Am. J. Physiol.*, 1901, v, 301–320.
419 PEARL, R.: The Movements and Reactions of Fresh-water *Planarians*: a Study in Animal Behavior. *Quart. J. Micr. Sc.*, 1902–03, xlvi, 509–714.
420 PEARL, R., and COLE, L. J.: The Effect of Very Intense Light on Organisms. *Third Rep. Mich. Acad. Sc.*, 1901, 77–78.
421 PEARSE, A. S.: The Reactions of Amphibians to Light. *Proc. Am. Acad. Arts and Sc.*, 1910, xlv, 161–208.
422 PÉREZ, J.: Notes zoologiques. De l'attraction exercée par les odeurs et les couleurs sur les insects. *Acta Soc. Linn., Bordeaux*, 1894, vii, 245–253.
423 PFEFFER, W.: Locomotorische Richtungsbewegungen durch chemische Reize. *Ber. bot. Ges.*, 1883, i, 524–533.
424 PFEFFER, W.: Locomotorische Richtungsbewegungen durch chemische Reize. *Unters. Bot. Inst. Tübingen*, 1884, i, 363–482.
425 PFEFFER, W.: Ueber chemotaktische Bewegungen von Bakterien, Flagellaten und Volvocineen. *Unters. Bot. Inst. Tübingen*, 1888, ii, 582–661.
426 PHIPPS, C. F.: An Experimental Study of the Behavior of Amphipods with Respect to Light Intensity, Direction of Rays, and Metabolism. *Biol. Bull.*, 1915, xxviii, 210–223.
427 PLATEAU, F.: Recherches sur la perception de la lumière par les myriopodes aveugles. *J. Anat. et Physiol.*, 1886, xxii.
428 PLATEAU, F.: Nouvelles recherches sur les rapports entre les insectes et les fleurs. *Mém. Soc. Zool. France*, 1899, xii.
429 PLATEAU, F.: La choix des couleurs par les insectes. *Mém. Soc. Zool. France*, 1899, xii, 336–370.
430 PLATEAU, F.: Expériences sur l'attraction des insectes par les étoffes colorées et les objets brillants. *Ann. Soc. Ent. Belgique*, 1900, xliv.

[431] PLATT, J. B.: On the Specific Gravity of *Spirostomum, Paramœcium,* and the Tadpole in Relation to the Problem of Geotaxis. *Am. Nat.,* 1899, xxxiii, 31–38.

[432] POLIMANTI, O.: Ueber eine beim Phototropismus des *Lasius niger* L. beobachtete Eigentümlichkeit. *Biol. Centr.,* 1911, xxxi, 222–224.

[433] POLIMANTI, O.: Sul reotropismo nelle larve dei batraci (*Bufo e Rana*). *Biol. Centr.,* 1915, xxxv, 36–39.

[434] PORODKO, TH. M.: Vergleichende Untersuchungen über die Tropismen. I. Das Wesen der chemotropen Erregung bei den Pflanzenwurzeln. *Ber. bot. Ges.,* 1912, xxx, 16–27.

[435] PORODKO, TH. M.: II. Thermotropismus der Pflanzenwurzeln. *Ber. bot. Ges.,* 1912, xxx, 305–313.

[436] PORODKO, TH. M.: IV. Die Gültigkeit des Energiemengengesetzes für den negativen Chemotropismus der Pflanzenwurzeln. *Ber. bot. Ges.,* 1913, xxxi, 88–94.

[437] PORODKO, TH. M.: V. Das mikroskopische Aussehen der tropistisch gereizten Planzenwurzeln. *Ber. bot. Ges.,* 1913, xxxi, 248–256.

[438] POWERS, E. B.: The Reactions of Crayfishes to Gradients of Dissolved Carbon Dioxide and Acetic and Hydrochloric Acids. *Biol. Bull.,* 1914, xxvii, 177–200.

[439] PRENTISS, C. W.: The Otocyst of Decapod Crustacea: Its Structure, Development, and Functions. *Bull. Mus. Comp. Zool.,* 1901, xxxvi, 165–251.

[440] PRINGSHEIM, E. G.: Die Reizbewegungen der Pflanzen. Berlin, 1912, viii+326.

[441] PRINGSHEIM, E. G.: Das Zustandekommen der taktischen Reaktionen. *Biol. Centr.,* 1912, xxxii, 337–365.

[442] PRZIBRAM, K.: Ueber die ungeordnete Bewegung niederer Tiere. *Arch. ges. Physiol.,* 1913, cliii, 401–405.

[443] PÜTTER, A.: Studien über Thigmotaxis bei Protisten. *Arch. Anat. u. Physiol., Physiol. Abt.,* 1900, *Suppl.,* 243–302.

[444] RÁDL, E.: Ueber den Phototropismus einiger Arthropoden. *Biol. Centr.,* 1901, xxi, 75–86.

[445] RÁDL, E.: Untersuchungen über die Lichtreaktion der Arthropoden. *Arch. ges. Physiol.,* 1901, lxxxvii, 418–466.

[446] RÁDL, E.: Ueber die Lichtreaktionen der Arthropoden auf der Drehscheibe. *Biol. Centr.,* 1902, xxii, 728–732.

[447] RÁDL, E.: Untersuchungen über den Phototropismus der Tiere. Leipzig, 1903, viii+188.

[448] RÁDL, E.: Ueber die Anziehung des Organismus durch das Licht. *Flora,* 1904, xciii, 167–178.

[449] RÁDL, E.: Einige Bemerkungen und Beobachtungen über den Phototropismus der Tiere. *Biol. Centr.,* 1906, xxvi, 677–690.

[450] RÉAUMUR: Mémoires pour servir a l'histoire des insectes. Paris, 1740.
[451] REESE, A. M.: Observations on the Reactions of *Cryptobranchus* and *Necturus* to Light and Heat. *Biol. Bull.*, 1906, xi, 93–99.
[452] RILEY, C. F. C.: Observations on the Ecology of Dragon-fly Nymphs: Reactions to Light and Contact. *Ann. Ent. Soc. Am.*, 1912, v, 273–292.
[453] ROMANES, G. J.: Animal Intelligence. New York, 1883, pp. 520.
[454] ROMANES, G. J.: Jelly-fish, Star-fish and Sea-urchins. New York, 1893, x+323.
[455] ROTHERT, W.: Ueber Heliotropismus. *Beitr. Biol. Pflanzen*, 1894, vii, 1.
[456] ROTHERT, W.: Beobachtungen und Betrachtungen über taktische Reizerscheinungen. *Flora*, 1901, lxxxviii, 371–421.
[457] ROUX, W.: Ueber die Selbstordnung (Cytotaxis) sich "berührender" Furchungszellen des Froscheies durch Zusammenfügung, Zellentrennung und Zellengleiten. *Arch. Entwcklngsmech.*, 1896, iii, 381–468.
[458] ROYCE, J.: Outlines of Psychology. New York, 1903, pp. 417.
[459] RUCHLÄDEW, N.: Untersuchungen zur Kritik der Methodik chemotaktischer Versuche und zur Biologie der Leukozyten. *Z. Biol.*, 1910, liv, 533–559.
[460] SCHÄFER, K. L.: Ueber den Drehschwindel bei den Tieren. *Z. Psychol. u. Physiol. Sinnesorg.*, 1891.
[461] SCHAEFFER, A. A.: Reactions of *Ameba* to Light and the Effect of Light on Feeding. *Biol. Bull.*, 1917, xxxii, 45–74.
[462] SCHMID, B.: Ueber den Heliotropismus von *Cereactis aurantiaca*. *Biol. Bull.*, 1911, xxxi, 538–539.
[462a] SCHNEIDER, G. H.: Der tierische Wille. Leipzig, 1880.
[462b] SCHNEIDER, K. C.: Tierpsychologisches Praktikum in Dialogform. Leipzig, 1912, pp. 719.
[462c] SCHNEIDER, K. C.: Vorlesungen über Tierpsychologie. Leipzig, 1909.
[463] SCHOENICHEN, W.: Die Empfindlichkeit der Nachtschmetterlinge gegen Lichtstrahlen. *Prometheus*, 1904, xvi, 29–30.
[464] SCHOUTEDEN, H.: Le phototropisme de *Daphnia magna* Straus (Crust.). *Ann. Soc. Ent. Belgique*, 1902, xlvi, 352–362.
[465] SHIBATA, K.: Studien über die Chemotaxis der *Isoëtes*-Spermatozoiden. *Jahrb. wiss. Bot.*, 1905, xli, 561–610.
[466] SHOHL, A. T.: Reactions of Earthworms to Hydroxyl Ions. *Am. J. Physiol.*, 1914, xxxiv, 384–404.

LITERATURE

[467] SMITH, A. C.: The Influence of Temperature, Odors, Light, and Contact on the Movements of the Earthworm. *Am. J. Physiol.*, 1902, vi, 459–486.

[468] SMITH, G.: The Effect of Pigment-migration on the Phototropism of *Gammarus annulatus* S. I. Smith. *Am. J. Physiol.*, 1905, xiii, 205–216.

[469] SOSNOWSKI, J.: Untersuchungen über die Veränderungen des Geotropismus bei *Paramæcium aurelia*. *Bull. Internat. Acad. Sc. Cracovie*, 1899, 130–136.

[470] STATKEWITSCH, P.: Ueber die Wirkung der Induktionschläge auf einige Ciliata. *Le Physiologiste Russe*, 1903, iii, 41–45.

[471] STATKEWITSCH, P.: Galvanotropismus und Galvanotaxis der Ciliata. *Z. allg. Physiol.*, 1904, iv, 296–332; 1905, v, 511–534; 1907, vi, 13–43.

[472] STRASBURGER, E.: Wirkung des Lichtes und der Wärme auf Schwärmsporen. *Jenaische Z. Naturwiss.*, 1878, (N.F.) xii, 551–625. Also separate, Jena, pp. 75.

[473] SZYMANSKI, J. S.: Ein Versuch, das Verhältnis zwischen modal verschiedenen Reizen in Zahlen auszudrücken. *Arch. ges. Physiol.*, 1911, cxxxviii, 457–486.

[474] SZYMANSKI, J. S.: Aenderung des Phototropismus bei Küchenschaben durch Erlernung. *Arch. ges. Physiol.*, 1912, cxliv, 132–134.

[475] SZYMANSKI, J. S.: Ein Beitrag zur Frage über tropische Fortbewegung. *Arch. ges. Physiol.*, 1913, cliv, 343–363.

[476] SZYMANSKI, J. S.: Methodisches zum Erforschen der Instinkte. *Biol. Centr.*, 1913, xxxiii, 260–264.

[477] v. TAPPEINER, H.: Die photodynamische Erscheinung (Sensibilisierung durch fluoreszierende Stoffe). *Ergeb. Physiol.*, 1909, viii, 698–741.

[478] TERRY, O. P.: Galvanotropism of *Volvox*. *Am. J. Physiol.*, 1906, xv, 235–243.

[479] TORELLE, E.: The Response of the Frog to Light. *Am. J. Physiol.*, 1903, ix, 466–488.

[480] TORREY, H. B.: On the Habits and Reactions of *Sagartia davisi*. *Biol. Bull.*, 1904, vi, 203–216.

[481] TORREY, H. B.: The Method of Trial and the Tropism Hypothesis. *Science*, 1907, xxvi, 313–323.

[482] TORREY, H. B.: Trials and Tropisms. *Science*, 1913, xxxvii, 873–876.

[483] TORREY, H. B.: Tropisms and Instinctive Activities. *Psychol. Bull.*, 1916, xiii, 297–308.

[484] TORREY, H. B., and HAYS, G. P.: The Rôle of Random Movements in the Orientation of *Porcellio scaber* to Light. *J. Animal Behav.*, 1914, iv, 110–120.

485 TOWLE, E. W.: A Study in the Heliotropism of *Cypridopsis*. *Am. J. Physiol.*, 1900, iii, 345–365.
486 TURNER, C. H.: An Experimental Investigation of an Apparent Reversal of the Responses to Light of the Roach (*Periplaneta orientalis* L.). *Biol. Bull.*, 1912, xxiii, 371–386.
487 v. UEXKÜLL, J.: Vergleichend-sinnesphysiologische Untersuchungen. II. Der Schatten als Reiz für *Centrostephanus longispinus*. *Z. Biol.*, 1897, xxxiv, 319–339.
488 v. UEXKÜLL, J.: Die Wirkung von Licht und Schatten auf die Seeigel. *Z. Biol.*, 1900, xl, 447–476.
489 v. UEXKÜLL, J.: Umwelt und Innenwelt der Tiere. Berlin, 1909, pp. 261.
490 ULEHLA, VL.: Ultramikroskopische Studien über Geisselbewegung. *Biol. Centr.*, 1911, xxxi, 645–654, 657–676, 689–705, 721–731.
491 VAN HERWERDEN, M. A.: Ueber die Perzeptionsfähigkeit des Daphnienauges für ultra-violette Strahlen. *Biol. Centr.*, 1914, xxxiv, 213–216.
492 VERWORN, M.: Psycho-physiologische Protistenstudien. Experimentelle Untersuchungen. Jena, 1889, viii+219.
493 VERWORN, M.: Die polare Erregung der Protisten durch den galvanischen Strom. *Arch. ges. Physiol.*, 1889, xlv, 1–36; 1890, xlvi, 267–303.
494 VERWORN, M.: Gleichgewicht und Otolithenorgan. Experimentelle Untersuchungen. *Arch. ges. Physiol.*, 1891, 1, 423–472.
495 VERWORN, M.: Untersuchungen über die polare Erregung der lebendigen Substanz durch den konstanten Strom. *Arch. ges. Physiol.*, 1896, lxii, 415–450.
496 VERWORN, M.: Die polare Erregung der lebendigen Substanz durch den konstanten Strom. *Arch. ges. Physiol.*, 1896, lxv, 47–62.
497 VERWORN, M.: General Physiology. New York, 1899.
498 VIEWEGER, TH.: Recherches sur la sensibilité des infusoires (alcaliooxytaxisme), les réflexes locomoteurs, l'action des sels. *Arch. Biol.*, 1912, xxvii, 723–799.
499 DE VRIES, H.: Ueber einige Ursachen der Richtung bilateralsymmetrischer Pflanzenteile. *Arb. bot. Inst. Würzburg*, 1872, i, 223.
500 DE VRIES, M. S.: Die phototropische Empfindlichkeit des Segerhafers bei extremen Temperaturen. *Ber. bot. Ges.*, 1913, xxxi, 233–237.
501 WAGER, H.: On the Effect of Gravity upon the Movements and Aggregation of *Euglena viridis* Ehrb., and Other Microörganisms. *Phil. Trans. Roy. Soc. London*, 1911, cci, (B), 333–390.
502 WALLENGREN, H.: Zur Kenntnis der Galvanotaxis. I. Die anodische Galvanotaxis. *Z. allg. Physiol.*, 1903, ii, 341–384.

LITERATURE

503 WALLENGREN, H.: II. Eine Analyse der Galvanotaxis bei *Spirostomum*. *Z. allg. Physiol.*, 1903, ii, 516–555.
504 WALLENGREN, H.: III. Die Entwirkung des konstanten Stromes auf die inneren Protoplasmabewegungen bei den Protozoen. *Z. allg. Physiol.*, 1904, iii, 22–32.
505 WALTER, H. E.: The Reactions of *Planarians* to Light. *J. Exp. Zool.*, 1907, v, 35–162.
506 WASHBURN, M. F.: The Animal Mind. New York, 1909, pp. 333.
507 WEIGERT, F: Die chemischen Wirkungen des Lichts. Stuttgart, 1911.
508 WHEELER, W. M.: Anemotropism and Other Tropisms in Insects. *Arch. Entwcklngsmech.*, 1899, viii, 373–381.
509 WHITMAN, C. O.: Animal Behavior. Woods Hole Biol. Lectures, Boston, 1899, 285–338.
510 DE WILDEMAN, E.: Sur le thermotaxisme des *Euglènes*. *Bull. Soc. Belg. Micros.*, 1894, xx, 245–258.
511 V. WIESNER, J.: Heliotropismus und Strahlengang. *Ber. bot. Ges.*, 1912, xxx, 235–245.
512 WILLEM, V.: La vision chez les gastropodes pulmonés. *Compt. rend. Acad. Sc.*, 1891, cxii, 247–248.
513 WILLEM, V.: Sur les perceptions dermatoptiques. *Bull. Sc. France et Belgique*, 1891, xxiii, 329–346.
514 WILSON, E. B.: The Heliotropism of Hydra. *Am. Nat.*, 1891, xxv, 413–433.
515 WODSEDALEK, J. E.: Phototactic Reactions and Their Reversal in the May-fly Nymphs *Heptagenia interpunctata* (Say.). *Biol. Bull.*, 1911, xxi, 265–271.
516 YERKES, R. M.: Reaction of Entomostraca to Stimulation by Light. I. *Am. J. Physiol.*, 1899, iii, 157–182.
517 YERKES, R. M.: II. Reactions of *Daphnia* and *Cypris*. *Am. J. Physiol.*, 1900, iv, 405–422.
518 YERKES, R. M.: A Study of the Reactions and the Reaction Time of the Medusa *Gonionemus murbachii* to Photic Stimuli. *Am. J. Physiol.*, 1903, ix, 279–307.
519 YERKES, R. M.: Reactions of *Daphnia pulex* to Light and Heat. *Mark Anniversary Vol.*, 1903, 361–377.
520 YERKES, R. M.: The Reaction Time of *Gonionemus murbachii* to Electric and Photic Stimuli. *Biol. Bull.*, 1904, vi, 84–95.
521 ZAGOROWSKI, P.: Die Thermotaxis der *Paramæcien*. *Z. Biol.*, 1914, lxv, 1–12.
522 ZELIONY, G. P.: Observations sur des chiens auxquels on a enlevé les hémisphères cérébraux. *Compt. rend. Soc. Biol.*, 1913, lxxiv, 707–708.

[523] BLASIUS, E., and SCHWEIZER, F.: Elektrotropismus und verwandte Erscheinungen. *Arch. ges. Physiol.*, 1893, liii, 493–543.
[524] NERNST, W., and BARRATT, J. O. W.: Ueber die elektrische Nervenreizung durch Wechselströme. *Z. Electrochem.*, 1904, x, 664–668.
[525] MOORE, A. R.: The Action of Strychnine on Certain Invertebrates. *J. Pharm. and Exp. Therap.*, 1916, ix, 167–169.
[526] LOEB, J.: The Chemical Basis of Regeneration and Geotropism. *Science*, 1917, xlvi, 115–118.
[527] BREUER, J.: Ueber den Galvanotropismus (Galvanotaxis bei Fischen. *Sitzngsb. Akad. Wiss. Wien, mathem.-naturw. Kl.*, 1905, cxiv, 27–56.
[528] BREUER, J., and KREIDL, A.: Ueber die scheinbare Drehung des Gesichtsfeldes, während der Einwirkung einer Centrifugalkraft. *Arch. ges. Physiol.*, 1898, lxx, 494–510.
[529] HERMANN, L., and MATTHIAS, F.: Der Galvanotropismus der Larven von *Rana temporaria* und der Fische. *Arch. ges. Physiol.*, 1894, lvii, 391–405.
[530] JENSEN, P.: Ueber den Geotropismus niederer Organismen. *Arch. ges. Physiol.*, 1893, liii, 428–480.
[531] CROZIER, W. J.: The Photic Sensitivity of *Balanoglossus*. *J. Exp. Zool.*, 1917, xxiv, 211–217.
[532] CLAPARÈDE, E.: Les tropismes devant la psychologie. *J. Psychol. u. Neurol.*, 1908, xiii, 150–160.
[533] NAGEL, W. A.: Experimentelle sinnesphysiologiche Untersuchungen an Coelenteraten. *Arch. ges. Physiol.*, 1894, lvii, 495–552.
[534] SCHNEIDER, G. H.: Der tierische Wille. Leipzig, 1880.
[535] SCHNEIDER, K. C.: Tierpsychologisches Praktikum in Dialogform. Leipzig, 1912, pp. 719.
[536] SCHNEIDER, K. C.: Vorlesungen über Tierpsychologie. Leipzig, 1909.
[537] MORGULIS, S.: The Auditory Reactions of the Dog Studied by the Pawlow Method. *J. Animal Behav.*, 1914, iv, 142–145.
[538] MORGULIS, S.: Pawlow's Theory of the Function of the Central Nervous System and a Digest of Some of the More Recent Contributions to This Subject from Pawlow's Laboratory. *J. Animal Behav.*, 1914, iv, 362–379.
[539] CRAIG, W.: The Voices of Pigeons Regarded as a Means of Social Control. *Am. J. Sociology*, 1908, xiv, 86–100.
[540] CRAIG, W.: Male Doves Reared in Isolation. *J. Animal Behav.*, 1914, iv, 121–133.
[541] MATULA, J.: Untersuchungen über die Funktionen des Zentralnervensystems bei Insekten. *Arch. ges. Physiol.*, 1911, cxxxviii, 388–456.

LITERATURE

[542] LOEB, J.: Influence of the Leaf upon Root Formation and Geotropic Curvature in the Stem of *Bryophyllum calcycinum* and the Possibility of a Hormone Theory of These Processes. *Bot. Gaz.,* 1917, lxiii, 25–50.

[543] LOEB, J.: Untersuchungen zur physiologischen Morphologie der Tiere. I. Heteromorphose. II. Organbildung und Wachstum. Würzburg, 1891–1892.

[544] LOEB, J.: The Chemical Mechanism of Regeneration. *Ann. Inst. Pasteur,* 1918, xxxii, 1–16.

[545] CRAIG, W.: Appetites and Aversions as Constituents of Instincts. *Biol. Bull.,* 1918, xxxiv, 91–107.

[546] KANDA, S.: Further Studies on the Geotropism of *Paramœcium caudatum. Biol. Bull.,* 1918, xxxiv, 108–119.

[547] LYON, E. P.: Note on the Geotropism of *Paramœcium. Biol. Bull.,* 1918, xxxiv, 120.

[548] McCLENDON, J. F.: Protozoan Studies. *J. Exp. Zool.,* 1909, vi, 265–283.

[549] McEWEN, R. S.: The Reactions to Light and to Gravity in *Drosophila* and its Mutants. *J. Exp. Zool.,* 1918, xxv, 49–106.

[550] PAYNE, F.: Forty-nine Generations in the Dark. *Biol. Bull.,* 1910, xviii, 188–190.

[551] PAYNE, F.: *Drosophila ampelophila* Loew Bred in the Dark for Sixty-nine Generations. *Biol. Bull.,* 1911, xxi, 297–301.

[552] MORGAN, C. L.: Animal Behavior. London, 1900.

[553] STEVENS, N. M.: Regeneration in *Antennularia. Arch. Entwcklngsmech.,* 1910, xxx, pt. 1, 1–7.

[554] MAXWELL, S. S.: On the Exciting Cause of Compensatory Movements. *Am. J. Physiol.,* 1911–12, xxix, 367–371.

INDEX

Æschna, 30
Aglaophenia, 138
Allen, 39
Amblystoma, 41, 53, 59
Ammophila, 170
Amphipyra, 135
Anelectrotonus, 32ff.
Anemotropism, 132
Antennularia antennina, 119, 125
Arbacia, 148 ff.
Arenicola, 106, 108, 109
Aristotelian viewpoint of animal conduct, 17, 18
Asymmetrical animals, 70 ff.
Avena sativa, 84, 105, 106, 117
"Avoiding reactions," 96
Axenfeld, D., 54

Bacterium termo, 140, 142
Balanus eburneus, 75, 108
 perforatus, 116
Bancroft, F. W., 41 ff. 62, 72, 74, 98
Barratt, J. O. W., 146, 147
Barrows, W. M., 153, 154
Bauer, V., 18
Bees, heliotropic reactions of, 103 ff. 159
Bert, P., 101, 102
Blaauw, A. H., 84, 104, 106, 117
Blasius, E., 32
Blowfly, 51, 76, 109
Bohn, G., 75, 82
Brain lesions in fish, 24 ff.
 in dogs, 27 ff.
 in *Æschna*, 30
Bruchmann, H., 142
Bryophyllum calycinum, 22, 120, 125, 137
Buddenbrock, W., 18
Budgett, S. P., 46
Buller, A. H. R., 141, 143, 148 ff
Bunsen-Roscoe law, 21, 83 ff., 99, 100, 137
Butler, S., 161

Catelectrotonus, 32 ff.
Centrifugal force, 125, 126

Chemotropism, 139 ff., 160
Chilomonas, 144, 145
Chlamydomonas pisiformis, 106, 109
Cineraria, 48, 164
Circus movements, fish, 24 ff., dogs, 27 ff., *Æschna*, 30 housefly, 54, *Ranatra*, 54, *Proctacanthus*, 60, 61, 72, *Euglena*, 72 ff., *Vanessa Antiopa*, 54
Color sensations, 100 ff.
Colpidium colpoda, 144
Compensatory motions, 126, 128 ff.
"Conditioned reflexes," 166 ff.
Craig, W., 168
Crayfish, 38
Cucumaria cucumis, 125
Cypridopsis, 116

Danais plexippus, 162
Daphnia, 88, 89, 92, 96, 101, 102, 104, 113 ff, 162, 171, 172
Delage, Y., 123, 124
Dewitz, J., 136, 149
Diaptomus, 114, 115
Dragon fly larva, 30
Drosophila, 111, 116, 117, 153

Eudendrium, 66, 73, 83, 85, 106
Euglena, 16, 45, 62, 70, 72 ff., 97 ff., 106, 109
Ewald, W. F., 85, 88, 104, 116

"Fertilizin," 149, 150
Flourens, P., 27
Forced movements, 24 ff.
Franz, V., 18
"Fright reactions," 96
v. Frisch, K.., 103, 104
Fundulus, 143, 157

Galileo, 18
Galvanotropism, 32 ff.
Gammarus, 113, 114, 116
Garrey, W. E., 33, 40, 41, 51 ff., 71 72, 132, 133
Gelasimus, 39, 124
Geotropism, 119, ff.

Glaucoma scintillans, 144
Gonium, 109
Graber, V., 47, 100, 104
Groom, T. T., 112

Hammond, J. H. Jr., 68
Harper, E. H., 154
Heliotropic machine, 68 ff.
Heliotropism, 47 ff.
Hering, E., 127
Hermann, L., 32
Hess, C., 102 ff.
Holmes, S. J., 51 ff., 73, 116

Instincts, 156 ff.
"Irritability," 39
Isoëtes, 141, 142

Jellyfish, 41, 42
Jennings, H. S., 73, 96 ff., 119, 125, 143 ff., 155
Jordon, H., 18

Kellogg, V. L., 158
Knight, 125
Kreidl, A., 124
Kupelwieser, H., 104

Lidforss, B., 142
Lillie, F., 149 ff., 156
Littorina, 75
Lizard, nystagmus in, 126, 129, 130
Lubbock, J., 47
Ludloff, K., 43
Lumbricus, 109
Lummer-Brodhun photometer, 90
Lycopodium, 142
Lyon, E. P., 22, 125, 128, 131

McEwen, R. S., 111, 116 ff.
Mach, E., 33
Magendie, 27
Magnus, R., 22
Marchantia, 142
Mast, S. O., 73, 99, 108, 109, 119
Matula, J., 30
Maxwell, S. S., 33 ff., 106, 108, 126, 135
Mayer, A. G., 162
Mazda lamp, 86
Memory images, 164 ff.
Mendelssohn, M., 155
Ménière's disease, 17, 110
Miessner, B. F., 68

Moore, A. R., 22, 112, 115, 117
Morgulis S., 166
Muscle tension, 20 ff.; after brain lesions in fish, 24 ff., dogs, 27 ff.; under influence of, galvanic current, 32 ff., one source of light, 47 ff., two sources of light, 75 ff., changes in intensity of light, 95 ff.

Nereis, 135, 150
Nernst, W., 46, 95
Nernst lamps, 76
Neurons, orientation of, 38 ff.
Northrop, J. H., 75, 89, 90
Nystagmus, 126, 129, 130

Oltmanns, F., 117

Palæmon, 124
Palæmonetes, 33 ff., 52, 59
Pandorina, 106, 109
Paramœcium, 43 ff., 97, 125, 143, 144, 164 ff., 155
Parker, G. H., 54, 75
Patten, B. M., 75 ff., 92
Pawlow, 165 ff.
Payne, F., 118
Pfeffer, W., 140 ff.
Phacus Triqueter, 109
Phrynosoma, 126, 129, 130
Phycomyces, 105, 117
Platyonichus, 124
Polygordius, 115
Polyorchis penicillata, 41, 42
Porthesia chrysorrhœa, 48, 116, 161
Proctacanthus, 55 ff.

Rádl, E., 54, 88, 128
Ranatra, 52 ff.
Reflexes, 21 ff., 166
Retina images, 127 ff.
Reversal of helitropism, 112 ff.
Rheotropism, 131 ff.
Robber fly, 55 ff., 72

Sachs, 101.
Salamander larvæ, galvanotropism of, 41
Schweizer, F., 32
Scyllium canicula, 24
Serpula, 95
Shark, forced movements in, 22, 24 ff.
Sherrington, 22

INDEX 209

Shibata, K, 141, 142
Shock movements, 97, 98
Shrimp, galvanotropism in, 34 ff.
Soule, C. G., 162
Spirillum undula, 140
Spirographis spallanzani, 63, 83
Spondylomorum, 109
Steinach, 156
Stereotropism, 134 ff., 157
Stevens, N. M., 119
Sticklebacks, 132
Strongylocentrotus purpuratus, 151
Stylonychia mytilus, 144
Symmetry relations of animal body, 19 ff.

v. Tappeiner, H., 117
Terry, O. P., 44

Thermotropism, 155
Towle, E. W., 116
Trachelomonas euchlora, 109
"Trial and error," 17, 73, 153, 154
Tubularia mesembryanthemum, 137

v. Uexküll, J., 18, 21, 22

Vanessa antiopa, 54
Verworn, M., 42
Vitalism, 18
Volvox, 44, 45, 62, 83, 117

Wasteneys, H., 86, 99, 106, 108, 143
Wave lengths, heliotropic efficiency of, 100 ff.
Weber's law, 78, 142, 143
Wheeler, W. M., 132
Whitman, 158, 168

A CATALOGUE OF SELECTED DOVER BOOKS
IN ALL FIELDS OF INTEREST

A CATALOGUE OF SELECTED DOVER BOOKS
IN ALL FIELDS OF INTEREST

AMERICA'S OLD MASTERS, James T. Flexner. Four men emerged unexpectedly from provincial 18th century America to leadership in European art: Benjamin West, J. S. Copley, C. R. Peale, Gilbert Stuart. Brilliant coverage of lives and contributions. Revised, 1967 edition. 69 plates. 365pp. of text.

21806-6 Paperbound $3.00

FIRST FLOWERS OF OUR WILDERNESS: AMERICAN PAINTING, THE COLONIAL PERIOD, James T. Flexner. Painters, and regional painting traditions from earliest Colonial times up to the emergence of Copley, West and Peale Sr., Foster, Gustavus Hesselius, Feke, John Smibert and many anonymous painters in the primitive manner. Engaging presentation, with 162 illustrations. xxii + 368pp.

22180-6 Paperbound $3.50

THE LIGHT OF DISTANT SKIES: AMERICAN PAINTING, 1760-1835, James T. Flexner. The great generation of early American painters goes to Europe to learn and to teach: West, Copley, Gilbert Stuart and others. Allston, Trumbull, Morse; also contemporary American painters—primitives, derivatives, academics—who remained in America. 102 illustrations. xiii + 306pp. 22179-2 Paperbound $3.00

A HISTORY OF THE RISE AND PROGRESS OF THE ARTS OF DESIGN IN THE UNITED STATES, William Dunlap. Much the richest mine of information on early American painters, sculptors, architects, engravers, miniaturists, etc. The only source of information for scores of artists, the major primary source for many others. Unabridged reprint of rare original 1834 edition, with new introduction by James T. Flexner, and 394 new illustrations. Edited by Rita Weiss. 6⅝ x 9⅝.

21695-0, 21696-9, 21697-7 Three volumes, Paperbound $13.50

EPOCHS OF CHINESE AND JAPANESE ART, Ernest F. Fenollosa. From primitive Chinese art to the 20th century, thorough history, explanation of every important art period and form, including Japanese woodcuts; main stress on China and Japan, but Tibet, Korea also included. Still unexcelled for its detailed, rich coverage of cultural background, aesthetic elements, diffusion studies, particularly of the historical period. 2nd, 1913 edition. 242 illustrations. lii + 439pp. of text.

20364-6, 20365-4 Two volumes, Paperbound $6.00

THE GENTLE ART OF MAKING ENEMIES, James A. M. Whistler. Greatest wit of his day deflates Oscar Wilde, Ruskin, Swinburne; strikes back at inane critics, exhibitions, art journalism; aesthetics of impressionist revolution in most striking form. Highly readable classic by great painter. Reproduction of edition designed by Whistler. Introduction by Alfred Werner. xxxvi + 334pp.

21875-9 Paperbound $2.50

CATALOGUE OF DOVER BOOKS

THE PRINCIPLES OF PSYCHOLOGY, William James. The famous long course, complete and unabridged. Stream of thought, time perception, memory, experimental methods—these are only some of the concerns of a work that was years ahead of its time and still valid, interesting, useful. 94 figures. Total of xviii + 1391pp.
20381-6, 20382-4 Two volumes, Paperbound $8.00

THE STRANGE STORY OF THE QUANTUM, Banesh Hoffmann. Non-mathematical but thorough explanation of work of Planck, Einstein, Bohr, Pauli, de Broglie, Schrödinger, Heisenberg, Dirac, Feynman, etc. No technical background needed. "Of books attempting such an account, this is the best," Henry Margenau, Yale. 40-page "Postscript 1959." xii + 285pp.
20518-5 Paperbound $2.00

THE RISE OF THE NEW PHYSICS, A. d'Abro. Most thorough explanation in print of central core of mathematical physics, both classical and modern; from Newton to Dirac and Heisenberg. Both history and exposition; philosophy of science, causality, explanations of higher mathematics, analytical mechanics, electromagnetism, thermodynamics, phase rule, special and general relativity, matrices. No higher mathematics needed to follow exposition, though treatment is elementary to intermediate in level. Recommended to serious student who wishes verbal understanding. 97 illustrations. xvii + 982pp.
20003-5, 20004-3 Two volumes, Paperbound $6.00

GREAT IDEAS OF OPERATIONS RESEARCH, Jagjit Singh. Easily followed non-technical explanation of mathematical tools, aims, results: statistics, linear programming, game theory, queueing theory, Monte Carlo simulation, etc. Uses only elementary mathematics. Many case studies, several analyzed in detail. Clarity, breadth make this excellent for specialist in another field who wishes background. 41 figures. x + 228pp.
21886-4 Paperbound $2.50

GREAT IDEAS OF MODERN MATHEMATICS: THEIR NATURE AND USE, Jagjit Singh. Internationally famous expositor, winner of Unesco's Kalinga Award for science popularization explains verbally such topics as differential equations, matrices, groups, sets, transformations, mathematical logic and other important modern mathematics, as well as use in physics, astrophysics, and similar fields. Superb exposition for layman, scientist in other areas. viii + 312pp.
20587-8 Paperbound $2.50

GREAT IDEAS IN INFORMATION THEORY, LANGUAGE AND CYBERNETICS, Jagjit Singh. The analog and digital computers, how they work, how they are like and unlike the human brain, the men who developed them, their future applications, computer terminology. An essential book for today, even for readers with little math. Some mathematical demonstrations included for more advanced readers. 118 figures. Tables. ix + 338pp.
21694-2 Paperbound $2.50

CHANCE, LUCK AND STATISTICS, Horace C. Levinson. Non-mathematical presentation of fundamentals of probability theory and science of statistics and their applications. Games of chance, betting odds, misuse of statistics, normal and skew distributions, birth rates, stock speculation, insurance. Enlarged edition. Formerly "The Science of Chance." xiii + 357pp.
21007-3 Paperbound $2.50

CATALOGUE OF DOVER BOOKS

AMERICAN FOOD AND GAME FISHES, David S. Jordan and Barton W. Evermann. Definitive source of information, detailed and accurate enough to enable the sportsman and nature lover to identify conclusively some 1,000 species and sub-species of North American fish, sought for food or sport. Coverage of range, physiology, habits, life history, food value. Best methods of capture, interest to the angler, advice on bait, fly-fishing, etc. 338 drawings and photographs. l + 574pp. $6\frac{5}{8}$ x $9\frac{3}{8}$.
22383-1 Paperbound $4.50

THE FROG BOOK, Mary C. Dickerson. Complete with extensive finding keys, over 300 photographs, and an introduction to the general biology of frogs and toads, this is the classic non-technical study of Northeastern and Central species. 58 species; 290 photographs and 16 color plates. xvii + 253pp.
21973-9 Paperbound $4.00

THE MOTH BOOK: A GUIDE TO THE MOTHS OF NORTH AMERICA, William J. Holland. Classical study, eagerly sought after and used for the past 60 years. Clear identification manual to more than 2,000 different moths, largest manual in existence. General information about moths, capturing, mounting, classifying, etc., followed by species by species descriptions. 263 illustrations plus 48 color plates show almost every species, full size. 1968 edition, preface, nomenclature changes by A. E. Brower. xxiv + 479pp. of text. $6\frac{1}{2}$ x $9\frac{1}{4}$.
21948-8 Paperbound $5.00

THE SEA-BEACH AT EBB-TIDE, Augusta Foote Arnold. Interested amateur can identify hundreds of marine plants and animals on coasts of North America; marine algae; seaweeds; squids; hermit crabs; horse shoe crabs; shrimps; corals; sea anemones; etc. Species descriptions cover: structure; food; reproductive cycle; size; shape; color; habitat; etc. Over 600 drawings. 85 plates. xii + 490pp.
21949-6 Paperbound $3.50

COMMON BIRD SONGS, Donald J. Borror. $33\frac{1}{3}$ 12-inch record presents songs of 60 important birds of the eastern United States. A thorough, serious record which provides several examples for each bird, showing different types of song, individual variations, etc. Inestimable identification aid for birdwatcher. 32-page booklet gives text about birds and songs, with illustration for each bird.
21829-5 Record, book, album. Monaural. $2.75

FADS AND FALLACIES IN THE NAME OF SCIENCE, Martin Gardner. Fair, witty appraisal of cranks and quacks of science: Atlantis, Lemuria, hollow earth, flat earth, Velikovsky, orgone energy, Dianetics, flying saucers, Bridey Murphy, food fads, medical fads, perpetual motion, etc. Formerly "In the Name of Science." x + 363pp.
20394-8 Paperbound $2.00

HOAXES, Curtis D. MacDougall. Exhaustive, unbelievably rich account of great hoaxes: Locke's moon hoax, Shakespearean forgeries, sea serpents, Loch Ness monster, Cardiff giant, John Wilkes Booth's mummy, Disumbrationist school of art, dozens more; also journalism, psychology of hoaxing. 54 illustrations. xi + 338pp.
20465-0 Paperbound $2.75

CATALOGUE OF DOVER BOOKS

How to Know the Wild Flowers, Mrs. William Starr Dana. This is the classical book of American wildflowers (of the Eastern and Central United States), used by hundreds of thousands. Covers over 500 species, arranged in extremely easy to use color and season groups. Full descriptions, much plant lore. This Dover edition is the fullest ever compiled, with tables of nomenclature changes. 174 full-page plates by M. Satterlee. xii + 418pp.   20332-8 Paperbound $2.75

Our Plant Friends and Foes, William Atherton DuPuy. History, economic importance, essential botanical information and peculiarities of 25 common forms of plant life are provided in this book in an entertaining and charming style. Covers food plants (potatoes, apples, beans, wheat, almonds, bananas, etc.), flowers (lily, tulip, etc.), trees (pine, oak, elm, etc.), weeds, poisonous mushrooms and vines, gourds, citrus fruits, cotton, the cactus family, and much more. 108 illustrations. xiv + 290pp.   22272-1 Paperbound $2.50

How to Know the Ferns, Frances T. Parsons. Classic survey of Eastern and Central ferns, arranged according to clear, simple identification key. Excellent introduction to greatly neglected nature area. 57 illustrations and 42 plates. xvi + 215pp.   20740-4 Paperbound $2.00

Manual of the Trees of North America, Charles S. Sargent. America's foremost dendrologist provides the definitive coverage of North American trees and tree-like shrubs. 717 species fully described and illustrated: exact distribution, down to township; full botanical description; economic importance; description of sub-species and races; habitat, growth data; similar material. Necessary to every serious student of tree-life. Nomenclature revised to present. Over 100 locating keys. 783 illustrations. lii + 934pp. 20277-1, 20278-X Two volumes, Paperbound $6.00

Our Northern Shrubs, Harriet L. Keeler. Fine non-technical reference work identifying more than 225 important shrubs of Eastern and Central United States and Canada. Full text covering botanical description, habitat, plant lore, is paralleled with 205 full-page photographs of flowering or fruiting plants. Nomenclature revised by Edward G. Voss. One of few works concerned with shrubs. 205 plates, 35 drawings. xxviii + 521pp.   21989-5 Paperbound $3.75

The Mushroom Handbook, Louis C. C. Krieger. Still the best popular handbook: full descriptions of 259 species, cross references to another 200. Extremely thorough text enables you to identify, know all about any mushroom you are likely to meet in eastern and central U. S. A.: habitat, luminescence, poisonous qualities, use, folklore, etc. 32 color plates show over 50 mushrooms, also 126 other illustrations. Finding keys. vii + 560pp.   21861-9 Paperbound $3.95

Handbook of Birds of Eastern North America, Frank M. Chapman. Still much the best single-volume guide to the birds of Eastern and Central United States. Very full coverage of 675 species, with descriptions, life habits, distribution, similar data. All descriptions keyed to two-page color chart. With this single volume the average birdwatcher needs no other books. 1931 revised edition. 195 illustrations. xxxvi + 581pp.   21489-3 Paperbound $4.50

CATALOGUE OF DOVER BOOKS

TWO LITTLE SAVAGES; BEING THE ADVENTURES OF TWO BOYS WHO LIVED AS INDIANS AND WHAT THEY LEARNED, Ernest Thompson Seton. Great classic of nature and boyhood provides a vast range of woodlore in most palatable form, a genuinely entertaining story. Two farm boys build a teepee in woods and live in it for a month, working out Indian solutions to living problems, star lore, birds and animals, plants, etc. 293 illustrations. vii + 286pp.

20985-7 Paperbound $2.50

PETER PIPER'S PRACTICAL PRINCIPLES OF PLAIN & PERFECT PRONUNCIATION. Alliterative jingles and tongue-twisters of surprising charm, that made their first appearance in America about 1830. Republished in full with the spirited woodcut illustrations from this earliest American edition. 32pp. 4½ x 6⅜.

22560-7 Paperbound $1.00

SCIENCE EXPERIMENTS AND AMUSEMENTS FOR CHILDREN, Charles Vivian. 73 easy experiments, requiring only materials found at home or easily available, such as candles, coins, steel wool, etc.; illustrate basic phenomena like vacuum, simple chemical reaction, etc. All safe. Modern, well-planned. Formerly *Science Games for Children*. 102 photos, numerous drawings. 96pp. 6⅛ x 9¼.

21856-2 Paperbound $1.25

AN INTRODUCTION TO CHESS MOVES AND TACTICS SIMPLY EXPLAINED, Leonard Barden. Informal intermediate introduction, quite strong in explaining reasons for moves. Covers basic material, tactics, important openings, traps, positional play in middle game, end game. Attempts to isolate patterns and recurrent configurations. Formerly *Chess*. 58 figures. 102pp. (USO) 21210-6 Paperbound $1.25

LASKER'S MANUAL OF CHESS, Dr. Emanuel Lasker. Lasker was not only one of the five great World Champions, he was also one of the ablest expositors, theorists, and analysts. In many ways, his Manual, permeated with his philosophy of battle, filled with keen insights, is one of the greatest works ever written on chess. Filled with analyzed games by the great players. A single-volume library that will profit almost any chess player, beginner or master. 308 diagrams. xli x 349pp.

20640-8 Paperbound $2.75

THE MASTER BOOK OF MATHEMATICAL RECREATIONS, Fred Schuh. In opinion of many the finest work ever prepared on mathematical puzzles, stunts, recreations; exhaustively thorough explanations of mathematics involved, analysis of effects, citation of puzzles and games. Mathematics involved is elementary. Translated by F. Göbel. 194 figures. xxiv + 430pp. 22134-2 Paperbound $3.00

MATHEMATICS, MAGIC AND MYSTERY, Martin Gardner. Puzzle editor for Scientific American explains mathematics behind various mystifying tricks: card tricks, stage "mind reading," coin and match tricks, counting out games, geometric dissections, etc. Probability sets, theory of numbers clearly explained. Also provides more than 400 tricks, guaranteed to work, that you can do. 135 illustrations. xii + 176pp.

20338-2 Paperbound $1.50

CATALOGUE OF DOVER BOOKS

LAST AND FIRST MEN AND STAR MAKER, TWO SCIENCE FICTION NOVELS, Olaf Stapledon. Greatest future histories in science fiction. In the first, human intelligence is the "hero," through strange paths of evolution, interplanetary invasions, incredible technologies, near extinctions and reemergences. Star Maker describes the quest of a band of star rovers for intelligence itself, through time and space: weird inhuman civilizations, crustacean minds, symbiotic worlds, etc. Complete, unabridged. v + 438pp. 21962-3 Paperbound $2.50

THREE PROPHETIC NOVELS, H. G. WELLS. Stages of a consistently planned future for mankind. *When the Sleeper Wakes,* and *A Story of the Days to Come,* anticipate *Brave New World* and *1984,* in the 21st Century; *The Time Machine,* only complete version in print, shows farther future and the end of mankind. All show Wells's greatest gifts as storyteller and novelist. Edited by E. F. Bleiler. x + 335pp. (USO) 20605-X Paperbound $2.50

THE DEVIL'S DICTIONARY, Ambrose Bierce. America's own Oscar Wilde—Ambrose Bierce—offers his barbed iconoclastic wisdom in over 1,000 definitions hailed by H. L. Mencken as "some of the most gorgeous witticisms in the English language." 145pp. 20487-1 Paperbound $1.25

MAX AND MORITZ, Wilhelm Busch. Great children's classic, father of comic strip, of two bad boys, Max and Moritz. Also Ker and Plunk (Plisch und Plumm), Cat and Mouse, Deceitful Henry, Ice-Peter, The Boy and the Pipe, and five other pieces. Original German, with English translation. Edited by H. Arthur Klein; translations by various hands and H. Arthur Klein. vi + 216pp.
20181-3 Paperbound $2.00

PIGS IS PIGS AND OTHER FAVORITES, Ellis Parker Butler. The title story is one of the best humor short stories, as Mike Flannery obfuscates biology and English. Also included, That Pup of Murchison's, The Great American Pie Company, and Perkins of Portland. 14 illustrations. v + 109pp. 21532-6 Paperbound $1.25

THE PETERKIN PAPERS, Lucretia P. Hale. It takes genius to be as stupidly mad as the Peterkins, as they decide to become wise, celebrate the "Fourth," keep a cow, and otherwise strain the resources of the Lady from Philadelphia. Basic book of American humor. 153 illustrations. 219pp. 20794-3 Paperbound $1.50

PERRAULT'S FAIRY TALES, translated by A. E. Johnson and S. R. Littlewood, with 34 full-page illustrations by Gustave Doré. All the original Perrault stories—Cinderella, Sleeping Beauty, Bluebeard, Little Red Riding Hood, Puss in Boots, Tom Thumb, etc.—with their witty verse morals and the magnificent illustrations of Doré. One of the five or six great books of European fairy tales. viii + 117pp. 8⅛ x 11. 22311-6 Paperbound $2.00

OLD HUNGARIAN FAIRY TALES, Baroness Orczy. Favorites translated and adapted by author of the *Scarlet Pimpernel.* Eight fairy tales include "The Suitors of Princess Fire-Fly," "The Twin Hunchbacks," "Mr. Cuttlefish's Love Story," and "The Enchanted Cat." This little volume of magic and adventure will captivate children as it has for generations. 90 drawings by Montagu Barstow. 96pp.
(USO) 22293-4 Paperbound $1.95

CATALOGUE OF DOVER BOOKS

POEMS OF ANNE BRADSTREET, edited with an introduction by Robert Hutchinson. A new selection of poems by America's first poet and perhaps the first significant woman poet in the English language. 48 poems display her development in works of considerable variety—love poems, domestic poems, religious meditations, formal elegies, "quaternions," etc. Notes, bibliography. viii + 222pp.

22160-1 Paperbound $2.00

THREE GOTHIC NOVELS: THE CASTLE OF OTRANTO BY HORACE WALPOLE; VATHEK BY WILLIAM BECKFORD; THE VAMPYRE BY JOHN POLIDORI, WITH FRAGMENT OF A NOVEL BY LORD BYRON, edited by E. F. Bleiler. The first Gothic novel, by Walpole; the finest Oriental tale in English, by Beckford; powerful Romantic supernatural story in versions by Polidori and Byron. All extremely important in history of literature; all still exciting, packed with supernatural thrills, ghosts, haunted castles, magic, etc. xl + 291pp.

21232-7 Paperbound $2.50

THE BEST TALES OF HOFFMANN, E. T. A. Hoffmann. 10 of Hoffmann's most important stories, in modern re-editings of standard translations: Nutcracker and the King of Mice, Signor Formica, Automata, The Sandman, Rath Krespel, The Golden Flowerpot, Master Martin the Cooper, The Mines of Falun, The King's Betrothed, A New Year's Eve Adventure. 7 illustrations by Hoffmann. Edited by E. F. Bleiler. xxxix + 419pp. 21793-0 Paperbound $3.00

GHOST AND HORROR STORIES OF AMBROSE BIERCE, Ambrose Bierce. 23 strikingly modern stories of the horrors latent in the human mind: The Eyes of the Panther, The Damned Thing, An Occurrence at Owl Creek Bridge, An Inhabitant of Carcosa, etc., plus the dream-essay, Visions of the Night. Edited by E. F. Bleiler. xxii + 199pp.

20767-6 Paperbound $1.50

BEST GHOST STORIES OF J. S. LEFANU, J. Sheridan LeFanu. Finest stories by Victorian master often considered greatest supernatural writer of all. Carmilla, Green Tea, The Haunted Baronet, The Familiar, and 12 others. Most never before available in the U. S. A. Edited by E. F. Bleiler. 8 illustrations from Victorian publications. xvii + 467pp. 20415-4 Paperbound $3.00

MATHEMATICAL FOUNDATIONS OF INFORMATION THEORY, A. I. Khinchin. Comprehensive introduction to work of Shannon, McMillan, Feinstein and Khinchin, placing these investigations on a rigorous mathematical basis. Covers entropy concept in probability theory, uniqueness theorem, Shannon's inequality, ergodic sources, the E property, martingale concept, noise, Feinstein's fundamental lemma, Shanon's first and second theorems. Translated by R. A. Silverman and M. D. Friedman. iii + 120pp. 60434-9 Paperbound $1.75

SEVEN SCIENCE FICTION NOVELS, H. G. Wells. The standard collection of the great novels. Complete, unabridged. *First Men in the Moon, Island of Dr. Moreau, War of the Worlds, Food of the Gods, Invisible Man, Time Machine, In the Days of the Comet.* Not only science fiction fans, but every educated person owes it to himself to read these novels. 1015pp. 20264-X Clothbound $5.00

CATALOGUE OF DOVER BOOKS

AGAINST THE GRAIN (A REBOURS), Joris K. Huysmans. Filled with weird images, evidences of a bizarre imagination, exotic experiments with hallucinatory drugs, rich tastes and smells and the diversions of its sybarite hero Duc Jean des Esseintes, this classic novel pushed 19th-century literary decadence to its limits. Full unabridged edition. Do not confuse this with abridged editions generally sold. Introduction by Havelock Ellis. xlix + 206pp. 22190-3 Paperbound $2.00

VARIORUM SHAKESPEARE: HAMLET. Edited by Horace H. Furness; a landmark of American scholarship. Exhaustive footnotes and appendices treat all doubtful words and phrases, as well as suggested critical emendations throughout the play's history. First volume contains editor's own text, collated with all Quartos and Folios. Second volume contains full first Quarto, translations of Shakespeare's sources (Belleforest, and Saxo Grammaticus), Der Bestrafte Brudermord, and many essays on critical and historical points of interest by major authorities of past and present. Includes details of staging and costuming over the years. By far the best edition available for serious students of Shakespeare. Total of xx + 905pp.
21004-9, 21005-7, 2 volumes, Paperbound $7.00

A LIFE OF WILLIAM SHAKESPEARE, Sir Sidney Lee. This is the standard life of Shakespeare, summarizing everything known about Shakespeare and his plays. Incredibly rich in material, broad in coverage, clear and judicious, it has served thousands as the best introduction to Shakespeare. 1931 edition. 9 plates. xxix + 792pp. (USO) 21967-4 Paperbound $3.75

MASTERS OF THE DRAMA, John Gassner. Most comprehensive history of the drama in print, covering every tradition from Greeks to modern Europe and America, including India, Far East, etc. Covers more than 800 dramatists, 2000 plays, with biographical material, plot summaries, theatre history, criticism, etc. "Best of its kind in English," *New Republic*. 77 illustrations. xxii + 890pp.
20100-7 Clothbound $8.50

THE EVOLUTION OF THE ENGLISH LANGUAGE, George McKnight. The growth of English, from the 14th century to the present. Unusual, non-technical account presents basic information in very interesting form: sound shifts, change in grammar and syntax, vocabulary growth, similar topics. Abundantly illustrated with quotations. Formerly *Modern English in the Making*. xii + 590pp.
21932-1 Paperbound $3.50

AN ETYMOLOGICAL DICTIONARY OF MODERN ENGLISH, Ernest Weekley. Fullest, richest work of its sort, by foremost British lexicographer. Detailed word histories, including many colloquial and archaic words; extensive quotations. Do not confuse this with the Concise Etymological Dictionary, which is much abridged. Total of xxvii + 830pp. 6½ x 9¼.
21873-2, 21874-0 Two volumes, Paperbound $6.00

FLATLAND: A ROMANCE OF MANY DIMENSIONS, E. A. Abbott. Classic of science-fiction explores ramifications of life in a two-dimensional world, and what happens when a three-dimensional being intrudes. Amusing reading, but also useful as introduction to thought about hyperspace. Introduction by Banesh Hoffmann. 16 illustrations. xx + 103pp. 20001-9 Paperbound $1.00

CATALOGUE OF DOVER BOOKS

JOHANN SEBASTIAN BACH, Philipp Spitta. One of the great classics of musicology, this definitive analysis of Bach's music (and life) has never been surpassed. Lucid, nontechnical analyses of hundreds of pieces (30 pages devoted to St. Matthew Passion, 26 to B Minor Mass). Also includes major analysis of 18th-century music. 450 musical examples. 40-page musical supplement. Total of xx + 1799pp.
(EUK) 22278-0, 22279-9 Two volumes, Clothbound $15.00

MOZART AND HIS PIANO CONCERTOS, Cuthbert Girdlestone. The only full-length study of an important area of Mozart's creativity. Provides detailed analyses of all 23 concertos, traces inspirational sources. 417 musical examples. Second edition. 509pp. (USO) 21271-8 Paperbound $3.50

THE PERFECT WAGNERITE: A COMMENTARY ON THE NIBLUNG'S RING, George Bernard Shaw. Brilliant and still relevant criticism in remarkable essays on Wagner's Ring cycle, Shaw's ideas on political and social ideology behind the plots, role of Leitmotifs, vocal requisites, etc. Prefaces. xxi + 136pp.
21707-8 Paperbound $1.50

DON GIOVANNI, W. A. Mozart. Complete libretto, modern English translation; biographies of composer and librettist; accounts of early performances and critical reaction. Lavishly illustrated. All the material you need to understand and appreciate this great work. Dover Opera Guide and Libretto Series; translated and introduced by Ellen Bleiler. 92 illustrations. 209pp.
21134-7 Paperbound $1.50

HIGH FIDELITY SYSTEMS: A LAYMAN'S GUIDE, Roy F. Allison. All the basic information you need for setting up your own audio system: high fidelity and stereo record players, tape records, F.M. Connections, adjusting tone arm, cartridge, checking needle alignment, positioning speakers, phasing speakers, adjusting hums, trouble-shooting, maintenance, and similar topics. Enlarged 1965 edition. More than 50 charts, diagrams, photos. iv + 91pp. 21514-8 Paperbound $1.25

REPRODUCTION OF SOUND, Edgar Villchur. Thorough coverage for laymen of high fidelity systems, reproducing systems in general, needles, amplifiers, preamps, loudspeakers, feedback, explaining physical background. "A rare talent for making technicalities vividly comprehensible," R. Darrell, *High Fidelity*. 69 figures. iv + 92pp. 21515-6 Paperbound $1.00

HEAR ME TALKIN' TO YA: THE STORY OF JAZZ AS TOLD BY THE MEN WHO MADE IT, Nat Shapiro and Nat Hentoff. Louis Armstrong, Fats Waller, Jo Jones, Clarence Williams, Billy Holiday, Duke Ellington, Jelly Roll Morton and dozens of other jazz greats tell how it was in Chicago's South Side, New Orleans, depression Harlem and the modern West Coast as jazz was born and grew. xvi + 429pp.
21726-4 Paperbound $2.50

FABLES OF AESOP, translated by Sir Roger L'Estrange. A reproduction of the very rare 1931 Paris edition; a selection of the most interesting fables, together with 50 imaginative drawings by Alexander Calder. v + 128pp. 6½x9¼.
21780-9 Paperbound $1.25

CATALOGUE OF DOVER BOOKS

THE ARCHITECTURE OF COUNTRY HOUSES, Andrew J. Downing. Together with Vaux's *Villas and Cottages* this is the basic book for Hudson River Gothic architecture of the middle Victorian period. Full, sound discussions of general aspects of housing, architecture, style, decoration, furnishing, together with scores of detailed house plans, illustrations of specific buildings, accompanied by full text. Perhaps the most influential single American architectural book. 1850 edition. Introduction by J. Stewart Johnson. 321 figures, 34 architectural designs. xvi + 560pp.
22003-6 Paperbound $4.00

LOST EXAMPLES OF COLONIAL ARCHITECTURE, John Mead Howells. Full-page photographs of buildings that have disappeared or been so altered as to be denatured, including many designed by major early American architects. 245 plates. xvii + 248pp. 7⅞ x 10¾. 21143-6 Paperbound $3.50

DOMESTIC ARCHITECTURE OF THE AMERICAN COLONIES AND OF THE EARLY REPUBLIC, Fiske Kimball. Foremost architect and restorer of Williamsburg and Monticello covers nearly 200 homes between 1620-1825. Architectural details, construction, style features, special fixtures, floor plans, etc. Generally considered finest work in its area. 219 illustrations of houses, doorways, windows, capital mantels. xx + 314pp. 7⅞ x 10¾. 21743-4 Paperbound $4.00

EARLY AMERICAN ROOMS: 1650-1858, edited by Russell Hawes Kettell. Tour of 12 rooms, each representative of a different era in American history and each furnished, decorated, designed and occupied in the style of the era. 72 plans and elevations, 8-page color section, etc., show fabrics, wall papers, arrangements, etc. Full descriptive text. xvii + 200pp. of text. 8⅜ x 11¼.
21633-0 Paperbound $5.00

THE FITZWILLIAM VIRGINAL BOOK, edited by J. Fuller Maitland and W. B. Squire. Full modern printing of famous early 17th-century ms. volume of 300 works by Morley, Byrd, Bull, Gibbons, etc. For piano or other modern keyboard instrument; easy to read format. xxxvi + 938pp. 8⅜ x 11.
21068-5, 21069-3 Two volumes, Paperbound $10.00

KEYBOARD MUSIC, Johann Sebastian Bach. Bach Gesellschaft edition. A rich selection of Bach's masterpieces for the harpsichord: the six English Suites, six French Suites, the six Partitas (Clavierübung part I), the Goldberg Variations (Clavierübung part IV), the fifteen Two-Part Inventions and the fifteen Three-Part Sinfonias. Clearly reproduced on large sheets with ample margins; eminently playable. vi + 312pp. 8⅛ x 11. 22360-4 Paperbound $5.00

THE MUSIC OF BACH: AN INTRODUCTION, Charles Sanford Terry. A fine, nontechnical introduction to Bach's music, both instrumental and vocal. Covers organ music, chamber music, passion music, other types. Analyzes themes, developments, innovations. x + 114pp. 21075-8 Paperbound $1.25

BEETHOVEN AND HIS NINE SYMPHONIES, Sir George Grove. Noted British musicologist provides best history, analysis, commentary on symphonies. Very thorough, rigorously accurate; necessary to both advanced student and amateur music lover. 436 musical passages. vii + 407 pp. 20334-4 Paperbound $2.75

CATALOGUE OF DOVER BOOKS

A HISTORY OF COSTUME, Carl Köhler. Definitive history, based on surviving pieces of clothing primarily, and paintings, statues, etc. secondarily. Highly readable text, supplemented by 594 illustrations of costumes of the ancient Mediterranean peoples, Greece and Rome, the Teutonic prehistoric period; costumes of the Middle Ages, Renaissance, Baroque, 18th and 19th centuries. Clear, measured patterns are provided for many clothing articles. Approach is practical throughout. Enlarged by Emma von Sichart. 464pp. 21030-8 Paperbound $3.50

ORIENTAL RUGS, ANTIQUE AND MODERN, Walter A. Hawley. A complete and authoritative treatise on the Oriental rug—where they are made, by whom and how, designs and symbols, characteristics in detail of the six major groups, how to distinguish them and how to buy them. Detailed technical data is provided on periods, weaves, warps, wefts, textures, sides, ends and knots, although no technical background is required for an understanding. 11 color plates, 80 halftones, 4 maps. vi + 320pp. 6⅛ x 9⅛. 22366-3 Paperbound $5.00

TEN BOOKS ON ARCHITECTURE, Vitruvius. By any standards the most important book on architecture ever written. Early Roman discussion of aesthetics of building, construction methods, orders, sites, and every other aspect of architecture has inspired, instructed architecture for about 2,000 years. Stands behind Palladio, Michelangelo, Bramante, Wren, countless others. Definitive Morris H. Morgan translation. 68 illustrations. xii + 331pp. 20645-9 Paperbound $3.50

THE FOUR BOOKS OF ARCHITECTURE, Andrea Palladio. Translated into every major Western European language in the two centuries following its publication in 1570, this has been one of the most influential books in the history of architecture. Complete reprint of the 1738 Isaac Ware edition. New introduction by Adolf Placzek, Columbia Univ. 216 plates. xxii + 110pp. of text. 9½ x 12¾.
21308-0 Clothbound $10.00

STICKS AND STONES: A STUDY OF AMERICAN ARCHITECTURE AND CIVILIZATION, Lewis Mumford. One of the great classics of American cultural history. American architecture from the medieval-inspired earliest forms to the early 20th century; evolution of structure and style, and reciprocal influences on environment. 21 photographic illustrations. 238pp. 20202-X Paperbound $2.00

THE AMERICAN BUILDER'S COMPANION, Asher Benjamin. The most widely used early 19th century architectural style and source book, for colonial up into Greek Revival periods. Extensive development of geometry of carpentering, construction of sashes, frames, doors, stairs; plans and elevations of domestic and other buildings. Hundreds of thousands of houses were built according to this book, now invaluable to historians, architects, restorers, etc. 1827 edition. 59 plates. 114pp. 7⅞ x 10¾.
22236-5 Paperbound $3.50

DUTCH HOUSES IN THE HUDSON VALLEY BEFORE 1776, Helen Wilkinson Reynolds. The standard survey of the Dutch colonial house and outbuildings, with constructional features, decoration, and local history associated with individual homesteads. Introduction by Franklin D. Roosevelt. Map. 150 illustrations. 469pp. 6⅝ x 9¼. 21469-9 Paperbound $4.00

CATALOGUE OF DOVER BOOKS

ALPHABETS AND ORNAMENTS, Ernst Lehner. Well-known pictorial source for decorative alphabets, script examples, cartouches, frames, decorative title pages, calligraphic initials, borders, similar material. 14th to 19th century, mostly European. Useful in almost any graphic arts designing, varied styles. 750 illustrations. 256pp. 7 x 10.   21905-4 Paperbound $4.00

PAINTING: A CREATIVE APPROACH, Norman Colquhoun. For the beginner simple guide provides an instructive approach to painting: major stumbling blocks for beginner; overcoming them, technical points; paints and pigments; oil painting; watercolor and other media and color. New section on "plastic" paints. Glossary. Formerly *Paint Your Own Pictures.* 221pp.   22000-1 Paperbound $1.75

THE ENJOYMENT AND USE OF COLOR, Walter Sargent. Explanation of the relations between colors themselves and between colors in nature and art, including hundreds of little-known facts about color values, intensities, effects of high and low illumination, complementary colors. Many practical hints for painters, references to great masters. 7 color plates, 29 illustrations. x + 274pp.
20944-X Paperbound $2.75

THE NOTEBOOKS OF LEONARDO DA VINCI, compiled and edited by Jean Paul Richter. 1566 extracts from original manuscripts reveal the full range of Leonardo's versatile genius: all his writings on painting, sculpture, architecture, anatomy, astronomy, geography, topography, physiology, mining, music, etc., in both Italian and English, with 186 plates of manuscript pages and more than 500 additional drawings. Includes studies for the Last Supper, the lost Sforza monument, and other works. Total of xlvii + 866pp. 7⅞ x 10¾.
22572-0, 22573-9 Two volumes, Paperbound $10.00

MONTGOMERY WARD CATALOGUE OF 1895. Tea gowns, yards of flannel and pillow-case lace, stereoscopes, books of gospel hymns, the New Improved Singer Sewing Machine, side saddles, milk skimmers, straight-edged razors, high-button shoes, spittoons, and on and on . . . listing some 25,000 items, practically all illustrated. Essential to the shoppers of the 1890's, it is our truest record of the spirit of the period. Unaltered reprint of Issue No. 57, Spring and Summer 1895. Introduction by Boris Emmet. Innumerable illustrations. xiii + 624pp. 8½ x 11⅝.
22377-9 Paperbound $6.95

THE CRYSTAL PALACE EXHIBITION ILLUSTRATED CATALOGUE (LONDON, 1851). One of the wonders of the modern world—the Crystal Palace Exhibition in which all the nations of the civilized world exhibited their achievements in the arts and sciences—presented in an equally important illustrated catalogue. More than 1700 items pictured with accompanying text—ceramics, textiles, cast-iron work, carpets, pianos, sleds, razors, wall-papers, billiard tables, beehives, silverware and hundreds of other artifacts—represent the focal point of Victorian culture in the Western World. Probably the largest collection of Victorian decorative art ever assembled—indispensable for antiquarians and designers. Unabridged republication of the Art-Journal Catalogue of the Great Exhibition of 1851, with all terminal essays. New introduction by John Gloag, F.S.A. xxxiv + 426pp. 9 x 12.
22503-8 Paperbound $4.50

CATALOGUE OF DOVER BOOKS

DESIGN BY ACCIDENT; A BOOK OF "ACCIDENTAL EFFECTS" FOR ARTISTS AND DESIGNERS, James F. O'Brien. Create your own unique, striking, imaginative effects by "controlled accident" interaction of materials: paints and lacquers, oil and water based paints, splatter, crackling materials, shatter, similar items. Everything you do will be different; first book on this limitless art, so useful to both fine artist and commercial artist. Full instructions. 192 plates showing "accidents," 8 in color. viii + 215pp. 8⅜ x 11¼. 21942-9 Paperbound $3.50

THE BOOK OF SIGNS, Rudolf Koch. Famed German type designer draws 493 beautiful symbols: religious, mystical, alchemical, imperial, property marks, runes, etc. Remarkable fusion of traditional and modern. Good for suggestions of timelessness, smartness, modernity. Text. vi + 104pp. 6⅛ x 9¼. 20162-7 Paperbound $1.25

HISTORY OF INDIAN AND INDONESIAN ART, Ananda K. Coomaraswamy. An unabridged republication of one of the finest books by a great scholar in Eastern art. Rich in descriptive material, history, social backgrounds; Sunga reliefs, Rajput paintings, Gupta temples, Burmese frescoes, textiles, jewelry, sculpture, etc. 400 photos. viii + 423pp. 6⅜ x 9¾. 21436-2 Paperbound $4.00

PRIMITIVE ART, Franz Boas. America's foremost anthropologist surveys textiles, ceramics, woodcarving, basketry, metalwork, etc.; patterns, technology, creation of symbols, style origins. All areas of world, but very full on Northwest Coast Indians. More than 350 illustrations of baskets, boxes, totem poles, weapons, etc. 378 pp. 20025-6 Paperbound $3.00

THE GENTLEMAN AND CABINET MAKER'S DIRECTOR, Thomas Chippendale. Full reprint (third edition, 1762) of most influential furniture book of all time, by master cabinetmaker. 200 plates, illustrating chairs, sofas, mirrors, tables, cabinets, plus 24 photographs of surviving pieces. Biographical introduction by N. Bienenstock. vi + 249pp. 9⅞ x 12¾. 21601-2 Paperbound $4.00

AMERICAN ANTIQUE FURNITURE, Edgar G. Miller, Jr. The basic coverage of all American furniture before 1840. Individual chapters cover type of furniture—clocks, tables, sideboards, etc.—chronologically, with inexhaustible wealth of data. More than 2100 photographs, all identified, commented on. Essential to all early American collectors. Introduction by H. E. Keyes. vi + 1106pp. 7⅞ x 10¾. 21599-7, 21600-4 Two volumes, Paperbound $11.00

PENNSYLVANIA DUTCH AMERICAN FOLK ART, Henry J. Kauffman. 279 photos, 28 drawings of tulipware, Fraktur script, painted tinware, toys, flowered furniture, quilts, samplers, hex signs, house interiors, etc. Full descriptive text. Excellent for tourist, rewarding for designer, collector. Map. 146pp. 7⅞ x 10¾. 21205-X Paperbound $2.50

EARLY NEW ENGLAND GRAVESTONE RUBBINGS, Edmund V. Gillon, Jr. 43 photographs, 226 carefully reproduced rubbings show heavily symbolic, sometimes macabre early gravestones, up to early 19th century. Remarkable early American primitive art, occasionally strikingly beautiful; always powerful. Text. xxvi + 207pp. 8⅜ x 11¼. 21380-3 Paperbound $3.50

CATALOGUE OF DOVER BOOKS

VISUAL ILLUSIONS: THEIR CAUSES, CHARACTERISTICS, AND APPLICATIONS, Matthew Luckiesh. Thorough description and discussion of optical illusion, geometric and perspective, particularly; size and shape distortions, illusions of color, of motion; natural illusions; use of illusion in art and magic, industry, etc. Most useful today with op art, also for classical art. Scores of effects illustrated. Introduction by William H. Ittleson. 100 illustrations. xxi + 252pp.
21530-X Paperbound $2.00

A HANDBOOK OF ANATOMY FOR ART STUDENTS, Arthur Thomson. Thorough, virtually exhaustive coverage of skeletal structure, musculature, etc. Full text, supplemented by anatomical diagrams and drawings and by photographs of undraped figures. Unique in its comparison of male and female forms, pointing out differences of contour, texture, form. 211 figures, 40 drawings, 86 photographs. xx + 459pp. 5 3/8 x 8 3/8.
21163-0 Paperbound $3.50

150 MASTERPIECES OF DRAWING, Selected by Anthony Toney. Full page reproductions of drawings from the early 16th to the end of the 18th century, all beautifully reproduced: Rembrandt, Michelangelo, Dürer, Fragonard, Urs, Graf, Wouwerman, many others. First-rate browsing book, model book for artists. xviii + 150pp. 8 3/8 x 11 1/4.
21032-4 Paperbound $2.50

THE LATER WORK OF AUBREY BEARDSLEY, Aubrey Beardsley. Exotic, erotic, ironic masterpieces in full maturity: Comedy Ballet, Venus and Tannhauser, Pierrot, Lysistrata, Rape of the Lock, Savoy material, Ali Baba, Volpone, etc. This material revolutionized the art world, and is still powerful, fresh, brilliant. With *The Early Work,* all Beardsley's finest work. 174 plates, 2 in color. xiv + 176pp. 8 1/8 x 11.
21817-1 Paperbound $3.00

DRAWINGS OF REMBRANDT, Rembrandt van Rijn. Complete reproduction of fabulously rare edition by Lippmann and Hofstede de Groot, completely reedited, updated, improved by Prof. Seymour Slive, Fogg Museum. Portraits, Biblical sketches, landscapes, Oriental types, nudes, episodes from classical mythology—All Rembrandt's fertile genius. Also selection of drawings by his pupils and followers. "Stunning volumes," *Saturday Review.* 550 illustrations. lxxviii + 552pp. 9 1/8 x 12 1/4.
21485-0, 21486-9 Two volumes, Paperbound $10.00

THE DISASTERS OF WAR, Francisco Goya. One of the masterpieces of Western civilization—83 etchings that record Goya's shattering, bitter reaction to the Napoleonic war that swept through Spain after the insurrection of 1808 and to war in general. Reprint of the first edition, with three additional plates from Boston's Museum of Fine Arts. All plates facsimile size. Introduction by Philip Hofer, Fogg Museum. v + 97pp. 9 3/8 x 8 1/4.
21872-4 Paperbound $2.00

GRAPHIC WORKS OF ODILON REDON. Largest collection of Redon's graphic works ever assembled: 172 lithographs, 28 etchings and engravings, 9 drawings. These include some of his most famous works. All the plates from *Odilon Redon: oeuvre graphique complet,* plus additional plates. New introduction and caption translations by Alfred Werner. 209 illustrations. xxvii + 209pp. 9 1/8 x 12 1/4.
21966-8 Paperbound $4.00

CATALOGUE OF DOVER BOOKS

THE RED FAIRY BOOK, Andrew Lang. Lang's color fairy books have long been children's favorites. This volume includes Rapunzel, Jack and the Bean-stalk and 35 other stories, familiar and unfamiliar. 4 plates, 93 illustrations x + 367pp.
21673-X Paperbound $2.50

THE BLUE FAIRY BOOK, Andrew Lang. Lang's tales come from all countries and all times. Here are 37 tales from Grimm, the Arabian Nights, Greek Mythology, and other fascinating sources. 8 plates, 130 illustrations. xi + 390pp.
21437-0 Paperbound $2.50

HOUSEHOLD STORIES BY THE BROTHERS GRIMM. Classic English-language edition of the well-known tales — Rumpelstiltskin, Snow White, Hansel and Gretel, The Twelve Brothers, Faithful John, Rapunzel, Tom Thumb (52 stories in all). Translated into simple, straightforward English by Lucy Crane. Ornamented with headpieces, vignettes, elaborate decorative initials and a dozen full-page illustrations by Walter Crane. x + 269pp.
21080-4 Paperbound $2.50

THE MERRY ADVENTURES OF ROBIN HOOD, Howard Pyle. The finest modern versions of the traditional ballads and tales about the great English outlaw. Howard Pyle's complete prose version, with every word, every illustration of the first edition. Do not confuse this facsimile of the original (1883) with modern editions that change text or illustrations. 23 plates plus many page decorations. xxii + 296pp.
22043-5 Paperbound $2.50

THE STORY OF KING ARTHUR AND HIS KNIGHTS, Howard Pyle. The finest children's version of the life of King Arthur; brilliantly retold by Pyle, with 48 of his most imaginative illustrations. xviii + 313pp. 6⅛ x 9¼.
21445-1 Paperbound $2.50

THE WONDERFUL WIZARD OF OZ, L. Frank Baum. America's finest children's book in facsimile of first edition with all Denslow illustrations in full color. The edition a child should have. Introduction by Martin Gardner. 23 color plates, scores of drawings. iv + 267pp.
20691-2 Paperbound $2.50

THE MARVELOUS LAND OF OZ, L. Frank Baum. The second Oz book, every bit as imaginative as the Wizard. The hero is a boy named Tip, but the Scarecrow and the Tin Woodman are back, as is the Oz magic. 16 color plates, 120 drawings by John R. Neill. 287pp.
20692-0 Paperbound $2.50

THE MAGICAL MONARCH OF MO, L. Frank Baum. Remarkable adventures in a land even stranger than Oz. The best of Baum's books not in the Oz series. 15 color plates and dozens of drawings by Frank Verbeck. xviii + 237pp.
21892-9 Paperbound $2.25

THE BAD CHILD'S BOOK OF BEASTS, MORE BEASTS FOR WORSE CHILDREN, A MORAL ALPHABET, Hilaire Belloc. Three complete humor classics in one volume. Be kind to the frog, and do not call him names . . . and 28 other whimsical animals. Familiar favorites and some not so well known. Illustrated by Basil Blackwell. 156pp.
(USO) 20749-8 Paperbound $1.50

CATALOGUE OF DOVER BOOKS

MATHEMATICAL PUZZLES FOR BEGINNERS AND ENTHUSIASTS, Geoffrey Mott-Smith. 189 puzzles from easy to difficult—involving arithmetic, logic, algebra, properties of digits, probability, etc.—for enjoyment and mental stimulus. Explanation of mathematical principles behind the puzzles. 135 illustrations. viii + 248pp.
20198-8 Paperbound $1.75

PAPER FOLDING FOR BEGINNERS, William D. Murray and Francis J. Rigney. Easiest book on the market, clearest instructions on making interesting, beautiful origami. Sail boats, cups, roosters, frogs that move legs, bonbon boxes, standing birds, etc. 40 projects; more than 275 diagrams and photographs. 94pp.
20713-7 Paperbound $1.00

TRICKS AND GAMES ON THE POOL TABLE, Fred Herrmann. 79 tricks and games—some solitaires, some for two or more players, some competitive games—to entertain you between formal games. Mystifying shots and throws, unusual caroms, tricks involving such props as cork, coins, a hat, etc. Formerly *Fun on the Pool Table*. 77 figures. 95pp. 21814-7 Paperbound $1.00

HAND SHADOWS TO BE THROWN UPON THE WALL: A SERIES OF NOVEL AND AMUSING FIGURES FORMED BY THE HAND, Henry Bursill. Delightful picturebook from great-grandfather's day shows how to make 18 different hand shadows: a bird that flies, duck that quacks, dog that wags his tail, camel, goose, deer, boy, turtle, etc. Only book of its sort. vi + 33pp. $6\frac{1}{2}$ x $9\frac{1}{4}$.   21779-5 Paperbound $1.00

WHITTLING AND WOODCARVING, E. J. Tangerman. 18th printing of best book on market. "If you can cut a potato you can carve" toys and puzzles, chains, chessmen, caricatures, masks, frames, woodcut blocks, surface patterns, much more. Information on tools, woods, techniques. Also goes into serious wood sculpture from Middle Ages to present, East and West. 464 photos, figures. x + 293pp.
20965-2 Paperbound $2.00

HISTORY OF PHILOSOPHY, Julián Marias. Possibly the clearest, most easily followed, best planned, most useful one-volume history of philosophy on the market; neither skimpy nor overfull. Full details on system of every major philosopher and dozens of less important thinkers from pre-Socratics up to Existentialism and later. Strong on many European figures usually omitted. Has gone through dozens of editions in Europe. 1966 edition, translated by Stanley Appelbaum and Clarence Strowbridge. xviii + 505pp. 21739-6 Paperbound $3.00

YOGA: A SCIENTIFIC EVALUATION, Kovoor T. Behanan. Scientific but non-technical study of physiological results of yoga exercises; done under auspices of Yale U. Relations to Indian thought, to psychoanalysis, etc. 16 photos. xxiii + 270pp.
20505-3 Paperbound $2.50

Prices subject to change without notice.
Available at your book dealer or write for free catalogue to Dept. GI, Dover Publications, Inc., 180 Varick St., N. Y., N. Y. 10014. Dover publishes more than 150 books each year on science, elementary and advanced mathematics, biology, music, art, literary history, social sciences and other areas.